ROGER E. KANET

East-West Monograph Series **6**

Technology and Change in East-West Relations

edited by F. Stephen Larrabee

Institute for East-West Security Studies

WESTVIEW PRESS * BOULDER, COLORADO

Copyright © 1988 by the Institute for East-West Security Studies, Inc., New York.

All rights reserved.

The Institute for East-West Security Studies does not take or encourage specific policy positions. It is committed to encouraging and facilitating discussions of important issues of concern to East and West. The views expressed in this book do not necessarily reflect the opinions of the Board of Directors, the officers or the staff of the Institute.

This publication is made possible through the generous support of the Ford Foundation.

Distributed by Westview Press
    Frederick A. Praeger, Publisher
    5500 Central Avenue
    Boulder, Colorado 80301

Library of Congress Cataloging-in-Publication Data

Technology and change in East-West relations.

  (East-West monograph series; 6)
  Includes bibliographies.
  1. East-West trade (1945–   )   2. Communist countries—Foreign economic relations.   3. Technological innovations—International cooperation.   4. Technology transfer.   I. Larrabee, F. Stephen.   II. Series.
HF4050.T43   1988   337′.09171′7   88-32829

ISBN   0-913449-08-3   (IEWSS)
          0-8133-0661-2   (Westview)

Printed in the United States of America

TABLE OF CONTENTS

Preface 1
Introduction 3
      F. Stephen Larrabee and Mary Albon

I. Technology and East-West Relations

1. The Technological Factors
Shaping East-West Relations 19
    Eugene Skolnikoff

2. Technology and Social Change:
East-West Comparisons 47
    Jerzy Wiatr

II. Technology, Arms and Disarmament

3. The Impact of Technology on Nuclear
Deterrence and Strategic Arms Control 63
    Joseph S. Nye, Jr.

4. Advanced Technology and European Security:
Conceptual Considerations 85
    Andrzej Karkoszka

5. The Impact of Military Technology on
Conventional Arms Control in Europe 113
    Joachim Krause

6. The Impact of Military Technology on
the Arms Race 131
    Marek Thee

III. Technology and Economics

7. The Impact of Technological Change
on East-West Economic Relations 147
    Michael Kaser

8. Economic Forces in the History of
East-West Relations 165
    Jozsef Bognar

9. The Impact of Technology on Communications,
Culture and Public Policy Issues 181
    Pekka Tarjanne and Mauri K. Elovainio

IV. Technology and Policy

10. **The Impact of Technology on the Future of European Security and Cooperation** 195
    Konrad Seitz

11. **Technology and Public Policy in East-West Relations** 199
    Thomas W. Simons

About the Authors 207

Preface

The challenges to international stability and security and the opportunities for international cooperation presented by rapid advances in science and technology during the postwar period are of increasing concern to both East and West. Technology has brought us both good and ill. Its increasingly rapid development and application have radically altered the living standards and expectations of populations worldwide. At the very time when the superpowers are liquidating their arsenals of intermediate-range forces and appear poised to bring East-West relations to a new stage of cooperation, ballistic missile and nuclear technologies are proliferating to other countries. Technology has rendered global war more deadly, yet perhaps less likely. Mass communication and information technology has made it impossible for dictators left and right to imprison, torture or kill their countrymen without the news reaching the rest of the world, yet other developments in technology have introduced deadly new options for guerrilla groups and terrorists. Despite the fact that East and West often take different approaches to the issue of technology and its assimilation, concern over technology's adverse effects—military, political, social, economic, ecological—may one day eclipse the primacy of the ideological struggle between the two social systems.

The impact of technology on the future of East-West relations was the subject of the Institute for East-West Security Studies Sixth Annual Conference held June 11 to 13, 1987, in Helsinki, Finland. Over 150 participants attended, including Nobel laureates, government officials, media representatives and specialists from twenty-four countries in both East and West. The Helsinki Conference was the first joint East-West assessment of the impact of current and future technology on our mutual security. It attempted to address the areas which would be most affected by technological advances: 1) East-West relations; 2) international trade, finance and economic development; 3) nuclear, conventional and chemical/biological weapons arms control and reductions; and 4) policy and society.

This volume is based on papers presented at the Helsinki Conference which address the crucial aspects of these issue areas. Contributors include eminent specialists from both East

1

and West who evaluate the impact technology has already had on East-West relations and international stability, and attempt to anticipate the implications of rapid technological advances for the future. The changing—and growing—role of technology in all facets of life is explored thoroughly and with great insight by our contributors. The authors' diverse backgrounds are reflected in their views, yet a broad consensus does develop in this book: technology, despite its potential for destructive use, can and should be employed to develop greater cooperation in all aspects of international relations. Responsible guidance and foresight are needed to keep technological policy on the track of increased cooperation and interdependence.

The Helsinki Conference was made possible through the generous support of the Finnish Ministry of Foreign Affairs, Nokia Oy, Neste, Wartsila and Rauma-Repola. Finnish banks including Kansallis-Osake-Pannki, Union Bank of Finland, Skopbank and the Savings Banks also supported the conference. The Ford Foundation generously sponsored the 1986–1987 Resident Fellowships of Andrzej Karkoszka (Poland) and Joachim Krause (FRG), who wrote their contributions to this volume while in residence in New York. The Institute gratefully acknowledges the support of the Ford Foundation for the publication of this book. I would like to thank Institute staff members F. Stephen Larrabee, Vice President and Director of Studies, the editor of the volume; Peter B. Kaufman, Publications Officer, who guided the essays through the process of publication; and especially Mary Albon, Program Officer, without whose skills this volume would not have been possible. Christina Lombardi assisted in the editing of the papers.

The Institute for East-West Security Studies is pleased to sponsor the publication of this East-West monograph as a contribution to discussions in East and West about technology and change.

John Edwin Mroz
President
Institute for East-West
  Security Studies
New York
November 1988

Introduction

F. STEPHEN LARRABEE
MARY ALBON

The Global Nature of Technology

In the period since World War II, the world has witnessed an unprecedented, rapid expansion of technology and scientific innovation. The impact of this latest wave of the scientific-technological revolution has already been felt, in varying degrees, in every nation on Earth. Agriculture and manufacturing, communication and transport, education and access to information—each of these areas has benefited from scientific innovations of the past few decades. Technological advances have improved the standard of living for millions. But at the same time, technology is an ambiguous, double-edged force; technological advances have also contributed to both the quantitative increase and the qualitative improvement of weapons of mass destruction, from the atomic bombs dropped on Hiroshima and Nagasaki to contemporary conventional arms and battlefield nuclear weapons. As such, high technology has a major impact on international security and stability.

In June 1987 in Helsinki, Finland, the Institute for East-West Security Studies gathered over 150 representatives of NATO, the Warsaw Pact and the neutral and nonaligned European nations for a unique discussion of "Technology and Change in East-West Relations" at its Sixth Annual Conference. This volume is based on a series of papers delivered at the conference by specialists from the United States and both Eastern and Western Europe. Technology and its implications for international security are global issues that affect the two blocs and all other nations. In presenting views from both East and West, as well as those outside of, yet affected by, the East-West relationship, this book continues the Institute's goal of fostering a substantive East-West dialogue on crucial issues of security.

Although technology has had an enormous impact on social change throughout history, such as during the industrial revolution in Britain and elsewhere, the current wave of technological advancement has brought sweeping changes more quickly than ever before, and on a global scale. Not all contributors to this volume agree on the speed or degree of social change brought on by technology, but there is a consensus that technological change does contribute to social change. As Jerzy Wiatr of Poland points out in his chapter, the impact of rapid technological progress is a problem for all nations, but the character of the challenge and the social implications differ from country to country. Not all nations are at the same level of economic, social and political development, or share the same political system, and thus the rapidity of technological, economic and social changes both complicates the solution of old problems stemming from industrialization and creates new difficulties.

Konrad Seitz of the FRG notes in his chapter that the current wave of technological innovation has created not only a scientific-technological revolution but also social transformation: industrial societies are moving on to become information societies. In the West, according to Seitz, the age of the masses and social democracy is ending. The East, in order to be a competitive international power—as Soviet General Secretary Mikhail Gorbachev seems to recognize—must not remain mired in the industrial age. Eugene Skolnikoff of the United States concurs with Seitz, saying that the East soon may be faced with a Western economic system that is both more integrated and less responsive to governmental policy—in other words, a streamlined, deregulated and highly competitive economy. The challenge this poses to the East is to modify its own economic structures to enable it to take advantage of the advances of technology. Such changes will inevitably have political consequences, creating pressure for political changes in socialist societies.

Technology is now beginning to have a real impact on the policy process as well, Skolnikoff asserts in his chapter. With advances in communications and transportation, there is greater public knowledge of policy events with increasingly short lag time. According to Skolnikoff, knowledge increases public involvement in the policy process, both at the local and the national level; the further implication is the influence of the public on national foreign policy.

One theme that underlies many of the chapters in this book is the question of technological determinism. Does technology have a momentum of its own? What influence do policy decisions have on shaping the development of technological innovations and their impact on societies, economies and political systems? In his examination of arms race dynamics in the nuclear age, Marek Thee of Norway maintains that technological developments do have an internal momentum, or "push power," that contributes, along with economic-military and bureaucratic-political factors, to the qualitative arms race between East and West. Skolnikoff concurs in part, arguing that, although technological development does not occur without human direction, it can be considered an independent variable: "the cumulative effects and dispersed sources of technology continually create new and usually unforeseen requirements for policy." It is these unpredicted, and perhaps unpredictable, effects of technological development that present the greatest challenge to both East and West—a challenge that both sides can best meet together.

Although technology has the potential for creating conflicts of interest between nations and political groupings, it also engenders possibilities for international cooperation, including East-West cooperation. There are both economic and political incentives to increased cooperation in a number of fields of science and technology. The large scale of some projects, such as nuclear fusion, high-energy physics and civilian use of outer space, calls for the combined scientific and financial resources of East and West to develop efficiently, Konrad Seitz asserts.

A key component of the scientific-technological revolution under way today is the expansion and improvement in information and communication technologies. In particular, as Skolnikoff, Seitz, and the Finnish analysts Pekka Tarjanne and Mauri Elovainio point out in their respective chapters, advances in communication satellites and television broadcasting have already "shrunk" the world, creating an increasingly large viewing audience that transcends national boundaries. These advances make it difficult for a nation to remain isolated, or to broadcast different versions of events for domestic and foreign consumption. Attempts to limit access to information also hinder a society's international competitiveness. Tarjanne and Elovainio note that technological changes, in telecommunications as well as other areas, have the potential to either unite or deepen the divide between Eastern and Western Europe. The East, whether through self-imposed separation or through an inability to keep pace with Western advances, could become technologically isolated. The consequences of such a situation, Tarjanne and Elovainio suggest, could be such that the potential deepening of the technological and economic gap might have an indirect effect on the development of European security policies and harm the prevailing spirit of cooperation.

Again, technology's role is ambiguous: it can promote either divisiveness or cooperation. Although the technology gap between East and West is significant, and has the potential to increase dramatically, it is not the only division stemming from technology. Technology could also create rifts within the West: for example, as Skolnikoff points out, the United States may be less willing to maintain its military commitment to Western Europe and Japan if the international trade balance does not improve—and innovative and efficient applications of high technology will help determine the balance. Likewise West European economic performance—also based in large part on technology—may affect its role in NATO. Finally, although beyond the scope of this book, the gap between North and South must not be ignored.

Technology, Arms and Disarmament

The force of technology and scientific advances has the potential to either undermine the arms control process or enhance its constraints. The determining factor is the perception of the role and purpose of arms control by all parties involved. As Joseph Nye of the United States points out in his chapter, during the 1970s arms control was largely an attempt to codify the existing strategic balance. Technology has introduced an important element of volatility into this equation, however. Weapons systems have become far more sophisticated and accurate in recent years through the application of technological innovations, which has created the potential for increased arms competition and destabilization of the military-political balance in Europe.

The introduction of multiple independently targetable re-entry vehicles (MIRVs) into the U.S. strategic nuclear inventory of intercontinental ballistic missiles (ICBMs) dramatically underscores this point. When first introduced in the late 1960s, MIRV technology was regarded by the United States as stabilizing: it allowed the United States to increase the number of warheads in its arsenal without increasing the number of launchers, thus enabling the United States to maintain a significant military advantage. Hence the United States resisted efforts to constrain MIRV technology in SALT I. This advantage, however, proved only temporary. When the Soviet Union began MIRVing its missiles a few years later, the United States, to its dismay, found that, rather than having an advantage, its own land-based missile force was becoming increasingly vulnerable. Had the United States been willing to forego this temporary advantage and constrain MIRV technology, later concerns about the "window of vulnerability" might have largely been avoided.

Marek Thee's chapter also draws attention to the problems for arms control posed by modernization. He argues that arms control must shift its focus from quantities to qualities of weapons and attempt to constrain military technology. The issue, in his view, "is not numbers but the unconstrained expansion of military technology." As Thee points out, "whenever the technological feasibility of new weapons systems is proved in the laboratories, their production and deployment can scarcely be halted." As weapons—both nuclear and conventional—become increasingly accurate, rapid and deadly, the need to redirect arms control toward a qualitative approach becomes greater. One way to begin the reorientation process, Thee suggests, would be to restrict funding to military research and development (R&D) and to begin converting it to productive civilian applications.

Recent technological advances have shifted the focus of military concepts and R&D to possibilities for enhancing strategic defense. As a result, President Ronald Reagan's Strategic Defense Initiative (SDI) and the concept of ballistic missile defense (BMD) have received great attention. Joseph Nye addresses this issue in his chapter. He argues that "perhaps the most important question for the next decade or two will be whether technology will begin to reverse the dominance of offense over defense that has thus far characterized the nuclear era."

Although partial defense and defense dominance might be made credible as a result of new technologies, Nye notes that constructing defenses that will enhance the three dimensions of stable deterrence—crisis stability, arms race stability and political stability—will be difficult. He enumerates a number of areas related to SDI that could benefit from technological advances, including surveillance, acquisition, tracking and kill assessment (SATKA); directed-energy weapons; kinetic-energy weapons; and supporting technologies and battle management. However, their potential effects on SDI are uncertain. In Nye's words:

> Perhaps the most difficult challenges to SDI over the long term lie not with the weapons or sensors, but in the support functions of space lift, power systems, data processing, and overall battle management. Technologies in these areas will be critical in determining the cost-effectiveness of various weapons or SATKA concepts, how much of the overall system can be space-based, and whether each layer of the defense can be managed coherently as part of a complex system which can be operated at high levels of readiness on extremely short notice.

In the near term, in a strictly technological sense, it is not likely that the key criteria set by the Reagan administration as guidelines for deploying a defensive system—survivability and marginal cost efficiency—will be attained. In addition, technological developments may not yield the desired results: innovations could also contribute to the offensive capabilities of the other side, thus changing the character of the threat and perhaps leading to an offense-defense race. Clearly, as Eugene Skolnikoff warns, there is no such thing as a "technological fix," or an easy, comprehensive solution—as SDI was presented—to a situation that cannot be resolved by technology alone. Maintaining the strategic balance entails political guidance.

In the sphere of nuclear arms control, the volatility of technology complicates the arms control process. "If one is to avoid the debilitating impact of technology on arms control, however," writes Joseph Nye, "one has to conceive of arms control less as a set of legal agreements which freeze a technology or military balance and more as a process of communication about the management of a dynamic balance." The 1987 U.S.-Soviet treaty eliminating all intermediate-range nuclear forces (INF) should be seen in this light.

In the aftermath of the INF Treaty, conventional weapons have received greater attention. These weapons will be the subject of new negotiations, the Conventional Stability Talks (CST), tentatively scheduled to begin in Vienna in the spring of 1989. There is common concern in both East and West that the CST should not share the fate of the Mutual and Balanced Force Reduction (MBFR) talks and bog down over disputes on data. Konrad Seitz cites the goal of the June 1987 Berlin declaration of the WTO Political Consultative Committee—reducing the forces and conventional arms of both sides in Europe to levels at which each alliance can guarantee its own defense but lacks the means to launch a large-scale offensive or surprise attack—as a significant step in the right direction. While important differences of approach continue to separate the two sides, there is a growing consensus between East and West about the basic goals and objectives of any new talks on conventional arms and equipment.

In his chapter, Andrzej Karkoszka of Poland expresses concern about the increasing offensive potential of conventional weapons in Europe, asserting that such an orientation is prone to preemption by the other side, and is thus destabilizing. He identifies a number of systems currently in various stages of development which he views as among the most destabilizing. These include stealth aircraft and cruise missiles, which shorten warning time; electronic warfare systems; and, in particular, long-range reconnaissance and strike systems and related command, control, communication and intelligence (C^3I).

To deal with the impact of new technologies on conventional weapons within the negotiating framework, Karkoszka proposes a new, "preventive" approach to arms control. Rather than merely limiting or canceling planned weapons systems, "preventive" measures would attempt to forestall further destabilizing advances in weaponry before they can be developed. Karkoszka argues, "the ensuing dramatic expansion of conventional forces requires equally dramatic measures. They should be executed in addition to—and not as a substitute for—those undertaken in the traditional approach." These new measures include:

> 1) negotiating qualities instead of numbers; 2) tradeoffs between relevant weapons, not just similar kinds of weapons; 3) functions of military forces instead of potentials; 4) discussion of acquisition plans and processes instead of already existing weapons systems; and 5) giving doctrines and structures an equal footing in negotiations with operational characteristics and deployments.

Despite the ambition of and likely political difficulties entailed by such an agenda, Karkoszka maintains that a gradual process, with the initial emphasis placed on the most threatening and destabilizing conventional systems, could be undertaken, with the aim being "a pace surpassing that of technology."

Underlying Karkoszka's arguments is the assumption that

there is an arms race in conventional weapons in Europe. He blames the military-technological drive for fueling this competition between NATO and the WTO:

> [N]ew military technology, introduced by one side in the name of stability, eventually contradicts this justification [of preserving stability through introducing new or modernized weapons] the moment the other side introduces similar counter-technology. The vicious repetition of this cycle perpetuates basic mistrust and fortifies the heavy reliance of European states on military guarantees of security.

In contrast, Joachim Krause of the FRG suggests in his chapter that such an argument is applicable in the nuclear sphere, but not in the conventional area. He defines an arms race as "offensively oriented military preparations in peacetime by competing powers [which] might set off a spiral of force build-ups on both sides, even when these military preparations arise from defensive political intentions," and maintains that the conventional arms competition in Europe does not fit this scenario. Rather, Krause demonstrates through force comparisons of NATO and WTO arsenals that the increase in armaments is far from symmetrical, in terms of both numbers and qualities, and asserts that the driving force is the offensive orientation of the Warsaw Pact. In his view, Western technological upgrades are purely responsive actions to Warsaw Pact initiatives. Thus Krause defines the contemporary situation in Europe as

> a competition in which one side constantly strives for superiority and improvement of its offensive capacities, in order to secure swift and sweeping victory in a war that is to be dominated offensively from the outset, while on the other side, a military coalition tries to retain forces which will be able to negate the first side's prospects for such a victory, without necessarily trying to match its military efforts quantitatively or qualitatively.

This basic difference in approach to the conventional military confrontation in Europe underlines the need to clarify definitions of military concepts, doctrines and terminology in order to enhance understanding of each other's key concerns and fears in both East and West, which will in turn facilitate greater cooperation in arms control and disarmament efforts.

Technology and Economics

Security, however, encompasses far more than military and political issues. The strength and security of a nation increasingly depends on the condition of its economy. The ability to develop and apply high technology is a decisive (although not singular) factor in determining a nation's economic performance.

This fact is underscored, as Hungarian economist Jozsef Bognar points out in his chapter, by recent developments in the Soviet Union, China and the United States. Whereas the USSR and China have openly declared a shift in priorities toward economics, the United States out of necessity, he argues, will be forced to devote increasing attention and resources to issues of economic security. This could have a major impact on the world economy. As Bognar notes,

> an economy-centered development model adopted by the world's three leading powers fundamentally changes the international economic climate, since the evolution of world economic cooperation has often been rendered difficult by the unfavorable climate caused by political animosities between them.

Economic security must be taken seriously as a component of both East-West relations and global security. Its impact on these larger spheres will only grow greater in the future.

The move toward economic reform in the socialist countries in Eastern Europe, expanded and deepened in the three and a half years since Mikhail Gorbachev's assumption of power in the Soviet Union, has underscored the importance of the economic dimension of security. Reform in the East in the economic sphere will inevitably have political implications. Even the Soviet Union, in the words of Jerzy Wiatr of Poland, has "arrived at the threshold of systemic transformations." Gorbachev's policy of *perestroika*, or restructuring, calls for reforms in both the economic and the political spheres. Wiatr also asserts that "there is no way the socialist countries can avoid economic and political reforms. It is only the strategy of reforms, their speed and depth, and consequently their long-term effects that remain at this stage of the game."

The need for economic reform in Eastern Europe and the Soviet Union stems in part from the technology gap between East and West. Michael Kaser of the United Kingdom enumerates systemic factors in the socialist societies of Eastern Europe behind the technology gap. As he points out, in the Soviet administrative structure, technological R&D is isolated from production facilities; very little research is conducted at specialized institutes and universities, despite the obvious source of know-how; and there are few incentives for innovation in technological R&D. In addition, risk avoidance is a high priority in production; maximization of production is more important than product quality or innovation. The non-market price system is also a disincentive to R&D innovation, and inertia is a major factor in the lack of application of proven technological innovations.

In Hungary, usually singled out as a model of reform in the socialist world, economic reform was undertaken within the context of relative political consensus. In contrast, according to Bognar, the Soviet Union and China first required a reorganization of political power. He points to the key role played by internal political conditions in the process of economic reform. Economic imperatives, he asserts, are neither isolated from political conditions nor sufficient in and of themselves to prompt reform.

The need for Eastern Europe and the Soviet Union to become integrated into the world economy was a pervasive theme at the Helsinki Conference. Pekka Tarjanne and Mauri Elovainio underscore this point in their discussion of the "technological renaissance" under way within the European Community (EC) in Western Europe. They see a gradual acceleration of the convergence of different technological sectors in the West, with information technologies being a crucial component in this process. This trend is likely to continue and intensify after the European Community's total economic integration goes into effect in 1992 and could contribute to increasing the gap—technological, but also socioeconomic—between East and West.

The East must integrate into the world economy to move away from the industrial society and toward the information society, Konrad Seitz argues. Eastern societies cannot develop all the technologies they need in isolation, and the Eastern market is not big enough to offset the costs of development and production of such technologies. In Seitz's view, integration of the East into the global economy would benefit world trade, which in turn would drive global economic growth.

One step toward integration would be the development of closer ties between the European Community and the Council of Mutual Economic Assistance (CMEA). Jozsef Bognar elaborates a series of changes that must be made within the socialist economies to facilitate their adaptation to and cooperation with the world economy: 1) shift toward an export orientation; 2) give priority to activities and cooperation directed at third (non-socialist) markets; 3) update methods of industrial and commercial cooperation with convertible-currency countries, and accept the need to export subcomponents and semi-finished products; 4) apply advanced technologies to improve the quality—and thus the competitiveness—of exports; 5) strive to increase imports of technology that further develop domestic production, services and infrastructure; 6) facilitate the establishment of joint ventures that provide needed technology and markets; 7) develop flexible trade policies; and 8) reorganize the linkage between production and distribution both within their own borders and for export.

Technology can either exacerbate current problems such as the gap between East and West or it can serve to help overcome them. Attempts at regulating the flow of some technologies deemed militarily applicable from the West to the East are becoming increasingly difficult—and costly—to enforce and are not necessarily effective. According to Eugene Skolnikoff, policies restricting technology transfer from West to East have had only limited success, and have entailed important costs for industry in the United States, political relations, and the open climate needed to maximize progress in R&D. Polish analyst Andrzej Karkoszka concurs with Skolnikoff, doubting the effectiveness of limitations on technology transfer for a number of reasons. Karkoszka points out that the Soviet defense industry—the major target of such policies—is quite powerful and not likely to be dependent on influxes of Western technology, including high technology. The concept of "dual-use" technologies (i.e., with both civilian and military applications) is obsolescent, Karkoszka argues, since most technologies are now dual-use. Finally, the expansion of Western restrictions on technology transfer to the East could further deepen the East-West division, as well as cause problems in relations between the United States and other Western nations.

In sharp contrast to the concerns about the transfer of sensitive technologies from West to East, Michael Kaser points out in his chapter that Eastern innovations from scientific and technological research do flow to the West. Indeed, the West usually has greater success in applying Eastern innovations than the societies in which they originate, largely because of the systemic problems in the East mentioned above.

Technology and Policy

A broad theme running through the Helsinki Conference was the impact of technology on the policy process. Because of developments in communication technologies, publics in both East and West are increasingly well informed about events taking place within their own countries and throughout the world. As Eugene Skolnikoff notes in his chapter, as a result, greater public knowledge and involvement in the policy process is inevitable. This is true not only regarding local and national issues, but increasingly for foreign policy as well, especially since the distinction between foreign and domestic policy has become blurred. In recent years, public opinion has become an ever more important factor in East-West relations.

Advances in technology may also have an impact on political choice. Not only are modern communication and information technologies extremely rapid, they are also increasingly difficult to control. With greater access to information (particularly about the policy process), publics are presented with alternative routes to dealing with problems and policy issues. They have the opportunity to compare their own political, social and economic systems to those of other states. Although there is greater openness and the speed of communication is faster in the West, ultimately the East must also experience the consequences of modern communication and increased access to information if it is to continue along the path of technological progress.

Several contributors to this volume see contemporary technological processes as leading to greater democracy in political life. Thomas Simons of the United States argues, for instance, that democracy is extremely competitive in the political arena, and the advantages of technology both reinforce and extend democracy's appeal. Similarly, Jerzy Wiatr maintains that, as a result of the latest wave of the scientific-technological revolution, democracy is a necessity. In his words: "technological progress without democracy is dangerous, since it will inevitably lead to technocratic deformations." At the same time he argues that without democracy—without the involvement of large social groups, particularly workers and the technical intelligentsia—it will be impossible to break down the barriers erected by conservative vested interests.

Simons also notes a tendency for technology to push politics toward "values" and away from specific issues. He writes that "technological development is destroying the power of traditional political organizations, along with their mass bases." New groups have the opportunity to take advantage of modern technologies, and benefit both economically and politically from this. Simons argues that the state, in both East and West, is undergoing a process of metamorphosis, and what remains is "a diffuse competition for diffuse support, and this can only be conducted as a competition about values." Modern telecommunications further contribute to this tendency in politics.

Technology is not a completely uncontrollable force, however. Eugene Skolnikoff emphasizes that technology does *not* predetermine policy, and that technological issues must be dealt with in political terms. As such, policy-makers have the ability to constrain and guide the course of technological expansion. Nevertheless, because technology is not an entirely independent variable in the policy arena, it can also be manipulated. This is particularly true in the case of the media. Yet technology's growing tendency toward allowing a pluralism of views to be presented to publics helps to counter the negative influence of efforts to control the flow of information.

Simons contends that the current trend in politics toward emphasizing values—and largely traditional values, he maintains—is not necessarily a destructive development. It will inject a new dose of competitiveness into politics and help keep the whole process viable. Simons asserts that since "debate over values, political competition, is an inescapable challenge for each of us, and to all of us together," the shared experience of this process will help bring East and West closer together.

This view ties in with Konrad Seitz's conviction that East and West must go beyond peaceful coexistence to develop cooperative interdependence to the benefit of all. East-West cooperation is needed to fight international problems facing all nations, such as pollution of the environment and deadly diseases like AIDS and cancer. Large-scale projects to combat international health and environmental threats call for cooperative endeavors because of both their costliness and the extent of their intended impact. This sort of cooperation will inevitably have far-reaching effects on East-West political relations. In the words of Seitz, "Peaceful coexistence of West and East is no longer enough; we need cooperative interdependence." Technology, with human guidance, can and should be a crucial element in the effort toward establishing a new, more cooperative stage in international relations.

I
Technology and East-West Relations

1

The Technological Factors Shaping East-West Relations

EUGENE B. SKOLNIKOFF

Technology is a pervasive and deceptive subject. It obviously has been a major factor in the transformation of society, and continues to be so. Yet, notwithstanding its apparent hard, factual character, rigorous analysis of its role in domestic or international affairs has not been easy. Anticipating the societal effects of future technological developments is even more difficult, though clearly of great importance. The difficulty stems in part from the unavoidable uncertainties of future technological change. But it also, even primarily, stems from the ubiquity of technology in the society, and the resulting challenge of separating technology from the multitude of other variables with which it interacts.

As a result, identifying technological trends and one by one exploring their likely impact in specific policy areas—seemingly the most obvious path for anticipating the effects of technology on East-West relations—could at best be only partially successful. It belies the intimate interaction of technology with politics, with economics, with social policy more generally, and even with other technology; it tends to give undue weight to new technologies as opposed to changing technological capabilities; it usually overvalues effects of dramatic technologies compared to cumulative effects of incremental change; and it denies the feedback from policy to the development of technology itself. To take but one example: an important element of East-West relations over the next decades will be the strength of China. But, that strength, in which technological capacity will be a significant element, will be determined not by the impact of new technology in China, but by the success of

China's policies toward the assimilation of technology and the development of its own capacity for innovation.

With this view in mind, major policy areas rather than technologies will be used in this chapter as a framework for exploring how technology is likely to shape the environment of East-West relations. Specific technological developments will certainly be relevant, but the intention is to see them in their social and issue setting in a way that illuminates more fundamental relationships. Some of the more important arenas of international affairs that are particularly relevant to East-West relations will be covered, arbitrarily divided into security and non-security issues. Such separation is not meant to imply that issues are independent of each other or that there is no feedback among them; quite the contrary in fact. A discussion of the effects of technology on policy processes is also included, because of their direct relevance to the conduct of relations between East and West. The time horizon is to the end of the century, with some attention to longer-term trends that could influence relations in the more immediate future.

Four generalizations about technology and its interaction with policy are worth noting as a preamble, since they are so often misunderstood or not recognized in consideration of policy, and are particularly relevant to the future (or any) policy environment.

1. Technological change does not lead to immediate and dramatic social change, as often portrayed in "futures" literature. Social change evolves through the impact of incremental developments in which technology is but one factor. Most new technologies, even seemingly radical ones, lead to social effects through an evolutionary process of social learning rather than sudden, quantum shifts. It may be arguable that the advent of atomic energy as a usable technology is one exception. Even if it is, there are not likely to be any others in the near future.

2. There is no such thing as a pure technological fix. That is, technology alone cannot be expected to solve an important societal problem without creating new social, economic, and political problems along the way, or in its wake. Technology can change the weight of the relevant factors, bring in new actors, and alter the costs and benefits. That is substantial; however, the societal consequences of technology—often unanticipated—mean that to turn to technology as an unen-

cumbered way to solve a serious political problem is a delusive goal.

3. As a corollary, all important technological issues in international politics must ultimately be dealt with in political terms. Technological factors are relevant, sometimes crucial, to understanding and dealing with many issues; but the policy choices of which technology is a part are not predetermined by technology. They will always finally turn on the political aspects.

4. Technological development does not happen independently of human direction. Though there is a sense in which technological change has a certain independent momentum, for policy purposes it is important to recognize that allocation choices in R&D *do* affect the characteristics of the resulting technology. At the same time, the cumulative effects and dispersed sources of technology continually create new and usually unforeseen requirements for policy. In effect, technology is both a dependent and an independent variable in the policy arena.

I. The Role of Technological Change in Economic and Political Relations

The overall effect of technology on major international political and economic (and, for that matter, security) relationships can be summed up by a shopworn, but still appropriate statement: the intensity of interaction and dependency among states, and among a larger number of states, will continue to increase, with technology a major factor—if not the major factor—making that possible and likely. That by now common observation, not less true because it is common, is a product of major developments in technology acting in concert with many other variables. Some of the more significant developments are those that modify constraints of time, distance, force, and cost as applied to human activities. Adding a new dimension to the effects is the institutionalization of technological development, and thus the certainty of continuing change.

The International Economy: Transactions, Competition and Economic Dependency

There are many dimensions to the interaction of technology with the international economy. Probably the most far-reaching is the enormous expansion in international transactions, trade, transborder data flows, movement of capital, and internationalization of business occurring in Western economies. Advances in communications and information technology have made this possible and profitable; the result has been major alteration of the international economic system, with much greater interlocking of economies and mutual dependence among them. This is not a completed change, but one that shows every sign of continuing, perhaps even accelerating. It is certain to affect the substance and the environment of East-West relations in many ways.

At the most general level, the continued expansion of international economic relationships that serves so intimately to couple Western economies will increase the pressure for coordination of national economic, fiscal and monetary policy in the West. Governments have tended to resist such coordination, fearing loss of domestic control; that response may be a luxury no longer possible, even for major economic powers. But, the same technological developments that lead to close coupling of economies and industry also encourage powerful transnational industrial relationships able to operate outside easy supervision by governments. The ability of Western nations to intervene decisively to control specific subjects such as the movement of technology, or to regulate the transborder flows of data and information, or even to achieve broader economic policy purposes, is thus likely to be even more uncertain than it is today. The East may well be faced with an economic system in the West that is both more integrated and less responsive to governmental policy.

The evolution of economic structure is so far predominantly a phenomenon of the private sector in the West; East European nations, however, will necessarily be directly affected. The changes of business organization and operation in the West that grow out of the use of information technologies will certainly be one of the more significant factors determining the scale and nature of trade and economic relations between East and West. Perhaps of greater importance, those changes will challenge the economies of Eastern Europe to modify their own industrial structure to meet the advantages that intensive use of the latest information technology provides. The ability of industries in the East to assimilate these technologies at the necessary scale and rate, and the freedom they are allowed to follow economic incentives in establishing cross-bloc economic ties, will go far to determine their economic performance. But success would no doubt require structural changes in Eastern economies that would carry with them other important economic, and ultimately political, changes in the domestic societies of the nations of Eastern Europe.

There are other economic trends at work more specifically technology-related than these broad changes. One of these is the fashionable concern with "competitiveness," in which a nation's competence in high technology is perceived to be the crucial determinant of a nation's economic performance. It is, of course, not simply strength that is relevant, but the ability of an economy to turn that strength into innovative products able to capture and keep markets. How well a nation can do this against growing competition from other countries will affect more than a nation's trade balance and its ability to sustain its standard of living. It will also affect a nation's capacity to sustain international commitments, to support expensive constantly-changing military forces, and to avoid unwanted or excessive dependence on the policies of other nations. The willingness of the U.S. to maintain the present level of commitment to Western Europe or to Japan may ultimately be put in jeopardy if its trade imbalance is not improved; in turn, relative performance in high technology will be a major factor in determining what the trade balance will be.

Two current issues with long-term consequences illustrate some aspects of the competitiveness problem. One is the movement of high-technology industries from one nation to another: the migration of the semiconductor industry from the U.S. to Japan being the high-visibility example of today.[1] The concern is that the loss of the semiconductor chip manufacturing industry, a crucial industry whose role in support of civilian and military industry is akin to that played by steel in the last century, would create unacceptable dependence on policies of another nation. It is not only the possibility of willful denial of technology that is at stake. Of greater importance over time is the likely erosion of the capability for downstream design and application in electronics and computers, and the possibility of permanent loss of technological leadership and knowledge to other countries, knowledge that will then be out of U.S. control. From the Soviet point of view, this may be per-

---

1. "Report of Defense Science Board Task Force on Defense Semiconductor Dependency," Office of Undersecretary of Defense for Acquisition, January 23, 1987.

ceived largely as an internal Western realignment, though offering some enticing future possibilities. The U.S. government sees it as much more threatening to the nation's future military strength, and is considering major programs to attempt to correct the situation.[2] The disagreement over semiconductor chips is also raising the specter of a major high-technology trade war between the U.S. and Japan, which would have profound political and economic effects were it to happen.

A second example is closely related: the importance West European nations accord to being competitive in high technology with the U.S. and Japan. Western Europe has continued to be strong in technological innovation, but has been relatively less successful in the application and commercialization of its innovations.[3] Recognition of this weakness has led to a host of national and international efforts in Europe designed to spur indigenous European technological development and diffusion. Its success in this endeavor, encouraging but by no means assured, will directly affect its economic performance, and through that the nature of Western Europe's role in the NATO alliance and in economic relations with Eastern Europe.

As far as the Soviet Union is concerned, there is little reason to expect appreciable narrowing of its high-technology gap with the West without extensive changes in economic structure. Mr. Gorbachev seems clearly to understand this, but the kind of changes required to improve substantially the Soviet Union's capacity for technological innovation—for example, decentralized decision-making, loosening of controls over information flow and computers, restructuring of industry to create incentives for innovation and to encourage independent initiative—will pose major, perhaps fundamental, challenges to the entire Soviet political system. If the changes were to be made, the political as well as economic climate for relations between East and West might be very different indeed. Without them, there is little reason to believe the Soviet Union's lag with respect to Western technology will decrease, or that its dependence on the West for technology will diminish. Note that this technological lag is largely independent of Western

---

2. Ibid.

3. Stanley Woods, *Western Europe: Technology and the Future*, Atlantic Paper No. 63 (London: Atlantic Institute for International Affairs, 1987).

attempts to deny advanced technology. U.S. programs to limit technology flow, discussed later, have in fact done more harm to U.S. industry than they have hampered the Soviet Union.[4]

The strategic military significance of this technological lag is not likely to be great. From an economic perspective, however, the inability to move efficiently to an information economy would greatly hamper the Soviet Union's performance relative to the West. As a result, its ability to maintain a dynamic global foreign policy, while meeting its mounting domestic and bloc obligations, would likely be adversely affected. Conceivably, such a development over time could have major political repercussions within the Soviet Union. At the least, its interest in East-West trade in technology would likely be even more evident than today. This interest will no doubt continue to be met with some hostility in the U.S., even if not necessarily as virulent as at present. Growing technological capability in third countries will, however, create a larger number of qualified trade partners for the Soviets, making the U.S. a less important technology supplier.

It is worth noting that technology will also be a factor in *reducing* the importance to industrialized countries of some traditional geopolitical concerns. For example, application of science and technology to agriculture has drastically altered the nature of food dependencies. Global food production, for many years to come, can be adequate to feed the world's population, though economic, political and institutional problems may interfere with the actual distribution of food to those who need it. The Soviet Union may still have to depend on food from outside its borders because of internal problems of production and distribution, but there need be no shortage of available food, and of willing suppliers. Resource dependencies in the future are also likely to be less important factors in international relationships, as technology makes it possible to find substitutes or bypass the need for scarce resources or to increase the efficiency of their use. The economic effects on nations dependent on export of mineral resources, however, will be severe; copper, for example, may never recover from the move to optical fiber transmission lines. Oil will be the

---

4. *Balancing the National Interest; U.S. National Security Export Controls and Global Economic Competition*, National Academy of Sciences, National Academy of Engineering, Institute of Medicine (Washington, DC: National Academy Press, 1987).

major exception to reduced resource dependencies, as technology has not yet been successfully applied to the development of economically suitable alternatives, and consumption in the West is increasing again. Clearly, that dependency on oil could once again become a major international issue, with the Soviet Union possibly benefiting financially and politically as an oil exporter in its dealings with the West. In a renewed "oil shock" climate, controversy over the Persian Gulf region could well erupt into more serious political or even military confrontations.

Multinational and Global Issues

One of the characteristic effects noted earlier of an increasingly technological world is to make interaction among nations more intense and at the same time unavoidable. This century has seen the application of technology in many fields such as health, space, weather, agriculture, pollution and others that has then required international or multinational action of some sort, either to deal with the effects of the technology, or to reap its benefits. These developments in "functional" subjects have brought about important cumulative changes on the international scene, but have only occasionally been seen as being at the heart of contemporary international politics, and especially not of East-West relations.

However, this may change as some of the issues become of much greater direct significance to the major industrial nations. In fact, in a few cases, there could be profound effects on the fate of all peoples and nations, with a possible requirement for cooperative action, especially among the nations of East and West, that goes beyond any previous experience. In particular, four issues stand out, though on quite different time scales. One is the danger of radioactive pollution as a result of nuclear accidents, the second the so-called greenhouse effect arising from carbon dioxide accumulation, the third the prospect of a worldwide AIDS epidemic, and the fourth the effects of population change. All deserve brief comment.

1. Radioactive Pollution

Environmental pollution across borders and globally is an issue that reached prominence on the international political scene at the time of the Stockholm Conference in 1972, though of course international environmental problems have been around much longer. None of the pollution issues of that time or since, including acid rain and Rhine river chemical spills, had as severe political repercussions as those caused by the 1986 Chernobyl accident. Its effects throughout Europe, the intense pressure on the Soviet Union to provide unprecedented amounts of information, and the political demand for international planning for possible future nuclear problems, have created a deep sense of the intimate linkage of Eastern and Western Europe in the face of new and dangerous technologies. The reaction was amplified by the pervasive psychological reaction to anything nuclear. Whether this accident and its aftermath will significantly affect political relations depends on many other factors, including whether there are any nuclear accidents in the near future, and what success anti-nuclear and other "green" political parties will have. It is bound to lead to some increased interaction and advance consultation on particularly sensitive environmental issues; whether it goes much beyond that politically will depend on the importance accorded to such issues in calculations of national interests.

2. The Greenhouse Effect

An environmental issue of a different kind is looming on the horizon, with major impact well into the next century, but with action required in the near future either to ameliorate that impact or to prepare for it. The issue is the so-called greenhouse effect which will result from the accumulation of CO_2 and other gases in the atmosphere, in time causing a change in the heat balance of the planet. These gases are the product of burning fossil fuel, of greater use of fertilizers, of increased animal husbandry, of deforestation; in other words, the product of a larger, wealthier and more industrialized global population. Climatic changes at the surface, when CO_2 in the atmosphere has doubled, are expected to be substantial, with predicted rises of 3°, +/- 1.5°C at the equator, and twice as much at the poles.[5] For comparison, this would be a larger temperature change

---

5. National Research Council, "Changing Climate: Report of the Carbon Dioxide Assessment Committee" (Washington, DC: National Academy Press, 1983).

than at any time since the ice ages. The time at which effects on this scale will occur is not certain, depending importantly on the rate of burning of fossil fuels and the contribution made by other gases, the latter now predicted to equal the effects of CO_2. 1985 estimates place the time of doubling as 50 to 100 years, that is between 2035 and 2085, but detectable effects could occur much earlier, even in the 1990s.

The specific consequences at the surface are not yet predictable in detail, but will surely involve major changes in precipitation, average temperature, cloud cover, sea level, and the frequency and severity of storms, droughts, and temperature extremes. In turn, these changes will, *inter alia*, alter the fertility of present agricultural areas and of some not now suitable for agriculture, lead to increased desertification, inundate some coastal areas, cause changes in food supply and availability, alter normal climate and weather patterns, and modify the animal and insect population. In short, the effects are likely to be substantial indeed. Contrary to some of the apocalyptic literature, however, it is not at all certain that on a global basis the effects will be catastrophic.[6] Food production worldwide might actually increase, and presently uninhabitable areas would become suitable for settlement.

What does seem clear at this time is that there will be winners and losers, and that some of the measures of power and influence, even of major nations, will be altered. Certainly, the economies of the nations of East and West will be affected, as the Soviet Union, for example, may ultimately benefit from greater access to and use of Siberian land and resources. New dependencies and interactions dictated by the new situation are likely to be created in ways that will affect East-West relations, and global politics generally, in the long run. The impact up to the turn of the century will grow out of greater consciousness of the magnitude of the phenomenon and its worldwide consequences. Substantial curtailment of fossil fuel consumption—the one step that would most affect the problem by delaying its arrival—is not a politically realistic option, though greater efficiency of use would help. But, it can be anticipated

---

6. There is one scenario that would be catastrophic: the loosening and sliding into the sea of the West Antarctic ice sheet, which would raise sea levels around the world about ten meters. This cannot be ruled out, but scientists now believe this would occur over a quite long time period, if it occurred at all.

that governments of major scientific countries will feel much increased pressure to engage in cooperative research on the phenomenon itself, and in cooperative planning for coping with its effects.

3. AIDS

Another issue normally removed from the political center of international affairs is health. The onset of a worldwide AIDS epidemic may, however, markedly alter that situation. It cannot be known for sure what the dimensions of the problem will be, but there is every possibility that AIDS will spread to all elements of society, and that the world will be faced with a disease that approximates the scope and impact of the bubonic plague; perhaps much worse.[7] It is even possible that mutations of the virus might make the disease more easily transmitted, and thus even more serious than now perceived. The only known hope to minimize the epidemic (other than draconian control of sexual relations) is progress in science and technology, and that can be hope only. If it were not for recent progress in molecular biology, there would be no reason even to hope for relief.

It is not at all beyond reality that AIDS, and the measures needed to cope with its effects, could become a major element in international affairs, and thus in East-West relations. National security issues will not be forgotten, and perhaps not directly affected at all. But, they may also be seen as less significant and more easily resolved in the face of the psychological stress of an unprecedented health crisis.

4. Population Change

Global population growth, aided and abetted by technology over the years, affects many issues that will increasingly influence the environment of East-West relations: migratory pressure, growing North-South economic disparities, and competition for land and water resources worldwide. However, one aspect of population change, made possible by technology,

---

7. Erik Eckholm, "AIDS, An Unknown Disease Before 1981, Grows Into A Worldwide Scourge," *The New York Times*, March 16, 1987, p. A17.

is often ignored even though it is likely to directly affect East-West relations. That aspect is the *decline* of population of many countries of Western Europe, and of the populations of European origin in the U.S. and USSR. The population of the Federal Republic of Germany, for example, is projected to decline by fully one third every generation (27 years), and the fertility rate in the U.S. is now below the "replacement" rate, with the population continuing to rise because of immigration.[8]

The effects of these changes will be felt largely after the turn of the century, with many problems for domestic policy: a growing proportion of elderly and particularly women, a reduced work force, expanded need for social services, and others. The international repercussions will also be substantial, however, for example in the availability of manpower for military service, changes in national economic performance, and possible changes in the relative priority given to international interests by electorates. The effects of these changes are little explored, but could affect East-West relations significantly early in the next century.

International Cooperation in Science and Technology

Large-scale international programs in science and technology may themselves change the landscape of international affairs, perhaps not in major ways, but enough to be more of a factor in East-West relations than they are today. Cooperative programs have been undertaken in the past for a variety of reasons, including the hope of scientific, economic and, at times, political gain. Those that bridge East and West have proven, not surprisingly, to be vulnerable to the changing political climate, whatever the other purposes that were being served. In fact, however, even in the dark days of political tension between the U.S. and the Soviet Union in the early 1980s, several U.S.-USSR cooperative agreements continued to operate.

The most significant change today is that the economic motive for cooperation is much more pressing in several scientific and technological areas, and may provide a more stable base for genuine cooperation. This is evident in high energy

---

8. Jane Menken, ed., *World Population and U.S. Policy; the Choices Ahead* (A Report of the American Assembly, Columbia University) (New York: W.W. Norton, 1986).

physics, in controlled magnetic fusion, possibly in space exploration, and in the fields relevant to the greenhouse effect and AIDS. The U.S. has announced a decision to proceed with a $6 billion accelerator—the Superconducting Supercollider—and is actively considering international cooperation as necessary to move to the next step in magnetic fusion research—a $4 billion test reactor. Both of these are almost certainly too large for the U.S. to carry out alone. Cooperation among Western countries is quite likely if the projects are to be done at all. Whether serious attempts will be made to engage the Soviet Union as a partner is yet to be decided, but the Soviet Union is strong in both of those fields and could participate as a rough equal if the political and other factors warranted. In fact, President Reagan and General Secretary Gorbachev announced stepped-up cooperation in magnetic fusion after their 1985 Geneva meeting.

Scientific and technological cooperation between East and West has had useful results in the past, though only occasionally has it been a major political factor in relations. Some have seen it as a route for undesirable transfer of militarily useful technology, while others have attempted to use cooperation as a policy instrument to influence unrelated issues. In fact, most careful reviews of East-West international cooperation indicate it can be valuable for the technical purposes in mind, and can serve broader modest political objectives.[9] The political changes that appear to be taking place in the Soviet Union, coupled with moderating attitudes in the U.S. that may change even faster after the 1988 election, could greatly enhance the prospect for expanded international cooperation in science and technology. If that happened, there might be several quite substantial, politically significant cooperative projects in the background of East-West relations within the decade.

Information/Communications

The "information revolution" is a cliché that, in fact, is an apt title for the continuing technological developments in com-

---

9. Carl Kaysen, Chairman, National Academy of Sciences, "Review of U.S.-USSR Interacademy Exchanges and Relations" (Washington, DC: 1977). Richard Garwin, Chairman, National Academy of Sciences, "Review of the U.S.-USSR Agreement on Cooperation in the Fields of Science and Technology" (Washington, DC: 1977).

puters and communications contributing to dramatic societal changes. The dimensions of that revolution are spelled out in many places and articles, though the striking continuing advances in those technologies—quite clearly far from leveling off—mean that their long-term meaning is not amenable to precise analysis. Their impact on the international economy has been discussed earlier. Another, perhaps more significant, impact has to do with their role in evolution of national culture and of political attitudes as the new technologies allow the massive intrusion of ideas and information from abroad, and make possible new modes of unsupervised interaction among individuals within and between societies. Policies that attempt to limit the flow of information, especially as direct broadcast satellites are deployed, and to control access to new computer technologies, are likely to become more difficult, and ultimately ineffective. They certainly would be incompatible with building a competitive modern society. The political implications for all countries, but particularly for those that have traditionally attempted to control information flows and to limit the use of technology, could be dramatic. The effects on the environment, and eventually the substance, of East-West relations would likely be considerable indeed, as better information becomes available to publics as well as elites, and as domestic political processes evolve under the new conditions.

Unfortunately, the reverse side of the coin also is feasible: the possibilities these technologies offer for influencing public opinion in organized ways that could directly impede, for example, improvements in relations between East and West. Or, of deeper concern, the possibility of using these information technologies to establish authoritarian control of a society or to attempt to undermine a society from abroad. This conflict among the various goals toward which technology can be applied, a typical and inevitable problem with most technologies, is certain to be characteristic of the applications of these technologies long into the future. There is no reason to expect, however, that the less desirable applications will dominate the effects that actually emerge. If they did, they would certainly not contribute to long-term improvement in East-West relations.

II. The Role of Technological Change in Security Relations

It would be difficult to overstate the role of technology in altering military-security issues, especially since World War II.

The direct effects on the massiveness of physical power, on the ability to exert force at a distance and all but instantaneously, on the complexity of weapons and decision-making, on capabilities for surveillance, and on others are all now familiar. These in turn have led to major change in the international political system: altering the meaning of resort to war, leading to the emergence of two competing nations dominating the global scene, and modifying the significance of many traditional geopolitical factors such as geography, forces in being, mobilization potential, population size, resource base and others.

The importance of technological change for security affairs gives every sign of continuing in the future, though the most important effects on East-West relations may lie only marginally in the direct impact on military weapons and hardware. As before in this analysis, the effects of general technological trends will be explored, rather than the impact of specific technological developments.

Reducing Differences among States: Proliferation, Conventional Weapons

In the perspective of the next decade or two, one of the more important effects of technology on weapons will be the diffusion of physical power to more and smaller countries. One possible route for that diffusion of power is proliferation of nuclear weapons. Though it is taking place at a slower pace than widely feared a decade ago, it is now possible for a substantial number of countries to master the technology and to obtain the necessary fissionable resources. Other factors than technology will determine whether any additional nations actually acquire nuclear weapons, but proliferation, or the evident possibility of proliferation, is likely to bedevil East-West relations for a long time to come. That prospect could serve to unite the superpowers in common policies, as it has in the past, or it could create serious conflicts and strains.

Another route for the diffusion of power is the continuing development of conventional weapons with increased firepower, greater accuracy, longer range, higher mobility and greater cost effectiveness. These weapons have not yet had as much R&D attention as more glamorous nuclear weapons, but even developments to date have the capability to alter the calculus of the non-nuclear battlefield. For example, stand-off offensive weapons; the families of weapons designed to destroy tanks, ships, aircraft and personnel; electronic battle control; and new capabilities for surveillance are gradually changing the characteristics, and the danger, of local war. The actual results of the changes are not precisely predictable; for example, superior Iraqi equipment has not been able to overcome Iranian zeal and scale of manpower, nor has Soviet technology been able to overcome local resistance in Afghanistan as the U.S. was also unable to do in Vietnam. Single missiles sank capital ships in the Falklands, and anti-aircraft missiles have severely modified operations requiring air cover in the Middle East.

Whatever the sometimes contradictory experience, it seems clear that technological development in general will give greater power and range to a small country, and thus the ability to extend its military reach and to require the greater use of force by those that might attack it. In short, "local" wars are likely to be more destructive, with greater possibility of escalation and threat of involvement of more countries. The concern over escalation of local conflict to involve the superpowers is likely to become more pressing as new technologies are developed and made available to a larger number of countries.

Developments in conventional technology will also have a direct effect on the NATO-Warsaw Pact confrontation, changing, for example, the significance of conventionally-tipped missiles that can now be extremely accurate at long ranges, or the viability of tanks or aircraft in a high-technology defense environment. Any actual or apparent changes in the conventional force balance between NATO and the Warsaw Pact will obviously have a direct impact on East-West relations. However, assessment of that balance today is subject to a high level of uncertainty and controversy, not likely to be resolved by the technological developments of the future. In fact, the rapid rate of change in technology-intensive weapons, and the lack of wartime experience, makes an assessment of the force balance particularly vulnerable to politically influenced analysis. That situation is not likely to improve, even when new technologies are put in place.

Lastly, the diffusion of physical power clearly will extend to non-state actors as well. The appearance on the arms market of new forms of easily transported high-power explosives and of formerly heavy and bulky long-range equipment that can now be carried by one man (such as anti-aircraft weapons) makes this certain. An upsurge of resistance movements or of terrorism, made more dangerous with high-technology weapons, could well become a factor in East-West relations, through, for example, destabilization of sensitive trouble-spots in which one or another superpower was heavily engaged.

Technology as Savior: SDI

The dramatic changes that have been wrought by rapid, and seemingly limitless, development of science and technology naturally have given rise to a belief that those fields can, under the right conditions, solve all problems. Experience, sometimes bitter, has shown how misplaced is that view. All real policy problems are as much or more non-technical in nature, as they are determined by science and technology. Even when a problem appears to be purely technical, science and technology can be only part of its resolution because of the relevance of non-technical factors at every stage of the R&D and application process.

Moreover, not all R&D goals are technically reasonable, even though it is not logically possible to prove that some developments *cannot* be realized. An unwise prediction by Vannevar Bush about the impossibility of intercontinental ballistic missiles is often cited. However, the weakness of a prediction that a particular piece of hardware with certain characteristics cannot be built is quite different from estimating the feasibility of an entire system, many of the requirements and problems of which are already known and which will have to operate in a hostile environment with constantly evolving technology. Even if the goals are theoretically "possible" (i.e., cannot be proven impossible), they may be simply foolish, akin to, for example, a project to send a spacecraft to land intact on the surface of the sun.

The world has been treated in the last few years to a throwback to the more innocent past. Science and technology have suddenly been put forward as being capable of solving the threat of nuclear war by creating a missile defense system that would make nuclear missiles impotent. The Strategic Defense Initiative (SDI) is a search for a technical fix to a situation that simply cannot be solved by technology alone. Even if some of the more exotic technologies visualized were to prove "feasible," the idea of creating a system of the cost and complexity contemplated—one that could never be demonstrated as a system on its own or against countermeasures, and that would have to live in an ever-changing, ever-threatening technological environment—is simply not a technological or political reality. Moreover, the information-handling and decision requirements of the system, even if soluble, would require response decisions in three minutes or less. It is difficult to imagine the political process in the U.S. resulting in a decision to accept a system carrying the distinct danger of initiating nuclear war without human intervention.

The weakness of the case has not prevented the commitment of substantial funds and the active promotion of the promise of SDI. Other motives are obviously at work on the part of many of its advocates. But whatever the motives, one can hazard the prediction today that the full conception of SDI will not for much longer be a serious objective. The current attempt to commit the United States to an early deployment decision is an act of desperation, not a recognition of accomplishment. Congress is unlikely to agree to deployment, or to vote the scale of R&D funds requested for the program. Attempts to reinterpret the ABM Treaty, as SDI testing would require, are likely to be resisted by Congress.

The goals of SDI may be cut back, but defensive technologies will continue to be explored by both sides, and will be a factor in future negotiations. Some form of ABM defense, probably ground-based, could in fact be a useful adjunct to agreed reductions in offensive arms.

The Soviet Union, for reasons of its own, has chosen to take very seriously the threat of SDI, perhaps because of the resource commitments required to hedge against uncertainty, and because of concern over the increased stimulation of innovative American science and technology. But the focus on SDI is likely to fade in the next few years as the U.S. moves away from a strong political commitment to it. Technological developments will continue, as in the recent past, to be incremental, to occur steadily but without causing strategic surprise, and to be roughly symmetrical over time in their effects on the strategic balance between the United States and the Soviet Union.

The Symbiosis of R&D, Security and Arms Control

The funds devoted to R&D worldwide have reached astonishing proportions. No accurate tabulation is available, but a rough calculation would indicate a total of some $400 billion per year.[10] Of that amount, a reasonable estimate, possibly conservative, is that one third is motivated directly or indirectly by military-security concerns. The scale of this commitment of resources is all the more striking when it is realized that in the United States in 1940, federal funds for defense R&D were $26.4 million, second to agriculture at $29.1 million.[11] Science and technology have become central aspects of national security concerns for nations of both East and West, and in a surprisingly short time.

This is not a novel observation, but from the East-West perspective there are several important implications often ignored in discussions of the specific products of R&D.

1. Change as a Constant

The fact that such a large proportion of the world's scientific and technological resources are devoted directly to national security, with much of the rest also contributing, means that the technological environment will never be static. There will be new capabilities, and frequent surprises, emerging from the laboratory or from industry. The current unexpected discovery of materials that are superconductive at higher temperatures than previously thought possible is an excellent example. That discovery will have many applications, including potentially important military applications. Other developments in fast-moving fields such as biotechnology are sure to have possible military uses as well.

But, these developments must be kept in perspective. They are effectively available, with only brief delays, to all sides. Moreover, there is no reason to expect that technical-military advances in any field visible today cannot be largely offset by parallel developments in countermeasures. This is particularly relevant to strategic systems of East and West, which, in effect, form one large interactive system, with substantial inertia and

---

10. Assuming roughly equal amounts in the U.S. and the Soviet Union, totaling approximately $240 billion, some $70 billion in Western Europe, $40 billion plus in Japan, and the remainder in the rest of the world.

11. Nathan Rosenberg, "Civilian Spillovers from Military R&D Spending: The American Experience Since World War II," Center for Economic Policy Research, Stanford University, September 1986.

little sensitivity to incremental change. New technological developments, even entirely new capabilities, are not likely to perturb the basic balance of the system to the end of the century at least, and probably well beyond.

The effects of new technological developments in the conventional arena could, however, be much more important, contributing, among other effects, to the trend noted above of making possible the diffusion of power to more and smaller states.

Large and sudden surprises with strategic effects remain extremely unlikely, but steady evolution of capabilities is certain. The details may be arguable, but the overall thrust is not. In effect, the results of science and technology will be constantly modifying the parameters of security relationships. The increasing number of countries with serious scientific and technological capabilities will in turn greatly complicate those relationships. Change, paradoxically, becomes a constant. The nations of East and West will be faced with this reality as they work with or confront each other in their myriad of security relationships.

2. Perceptions, Calculations of Power

Scientific and technological capability clearly is now a major ingredient of the relative power of states. But how is it to be measured? How important are perceptions of scientific and technological competence when so much of the determination of relative military capability is itself a matter of perception? There can be no definitive answer. It is clear, however, that states will have to be as concerned with public awareness of scientific and technological strength, as they are with any other aspect of their military posture. Both the Soviet Union and the United States appear to understand this today, with the perception of the lead held by the U.S. in science and technology partly reflected in the Soviet Union's extraordinary reaction to SDI. Any change in perception would have important effects on the climate of their relationship, as Sputnik had in the late 1950s.

3. Subversion of Arms Control

A corollary of constant change through R&D is that R&D is always destined to be "subversive" of arms control agreements.

That does not mean that such agreements cannot be in the security interests of states, or that R&D is not also important for development of the technology needed for implementation of arms control agreements. It simply means that negotiations have to be conducted in the knowledge that continued R&D is likely over time to change the elements and significance of an agreement, possibly leading to technology that makes an agreement obsolete. Especially is this so because nations will tend to focus research and development on weapons options that are not prohibited by arms control agreements.

Some agreements may attempt to control the significance of continued R&D, for example by prohibiting testing, but as the current dispute over the ABM Treaty shows, new technical possibilities may lead to new interpretations. Agreements designed to ban R&D directly are not desirable, and would be exceedingly hard to enforce, or even to visualize. Over the next decades, therefore, R&D itself will not likely be a direct focus of arms control negotiations, though it may be influenced by agreements on downstream technology. R&D will continue to be a factor that leads to evolution of the meaning of agreements after they are reached, and that leads to new capabilities requiring new methods of control.

4. Control of Technology Flows

The West's general lead in technology, and its significance for national security has led to severe attempts among OECD countries, and especially the United States, to limit the flow of technology to the East. These technology control policies have been a source of much controversy between the United States and its allies, and in fact within the U.S. itself. The expanded policies have been of only limited success, with important costs to industry in the United States, to political relations, and to the open climate necessary for maximum R&D progress.[12] Conversely, there is little doubt that the USSR and the Warsaw Pact countries have been anxious to obtain whatever technology they can from the West, and have resorted to a variety of legal and illegal means to do so. But the ability to control the transfer of all but the most critical technology is limited and costly, and the rise of technological capability in third countries makes

---

12. NAS report, *Balancing the National Interest* (see fn. 4).

technology that is outside the range of controls more readily available.

The United States, in recognition of the weakness and costs of control policies, is apparently planning to cut back controls so as to focus primarily on restricting new technology that is most directly relevant to military application and that is not readily available elsewhere. If actually implemented, this would allow the Soviet Union and its allies to have broader open access to Western technology in general, but would make it easier to protect what is most important, and to reduce the undesirable effects of the controls on R&D and on political relationships in the West. There is no reason to expect any changes in military capability to result from the eased technology transfer that would take place, but there could be potential for expanded trade, particularly between Eastern and Western Europe.

New Powers

The role of technology as a critical factor in both the economic and military strength of nations raises the question of the emergence of new powers on the international scene, strong enough to affect East-West relations over the next few decades. There are, in fact, several candidates. Japan stands out, obviously, as a technological-economic power that may possibly become a power in military terms as well. The United States has been pressing Japan to commit more funds to its military establishment, and to take more of a role in Asian defense. That nation has the human, economic and intellectual resources to play a larger strategic role; sooner or later it will, especially if trade relations with the United States seriously worsen, leading the U.S. to pull back its security guarantee.

China also may be poised on the edge of an economic takeoff similar to its Asian neighbor twenty years ago.[13] When that colossal and talented country does acquire modern technological capability on a nationwide scale it will certainly enter the lists as a major factor in world security affairs, and eventually as a superpower. That latter capability may still be some time off, but barring a return to the internal chaos of the 1960s, China's

---

13. Dwight H. Perkins, *China, Asia's Next Economic Giant?* (Seattle: University of Washington Press, 1986).

influence in economic and strategic matters is likely to grow steadily in the near future.

Other candidates are in the category commonly called Newly Industrializing Countries today. Brazil, South Korea and India are examples. None are likely to become major factors on the strategic stage unless they become embroiled in dangerous local conflict, but all are likely to be important economically and technologically, with the potential for eventual substantial military significance.

III. The Role of Technological Change in the Policy Process and Institutions

In considering the impact of technological change on international politics in general and East-West relations in particular, it is easy to neglect the likely changes in policy processes and institutions, both domestic and international. Those changes will be of great importance in the future, as they have been in the past, in establishing the framework in which relations will be considered, and in defining the bounds of possible policies. In keeping with the rest of this paper, the intent will be to try to single out underlying trends, rather than specific changes, recognizing, as always, that the trends are a product of technology working with other factors.

Public Involvement

In the West, technology has been an important factor in leading to much greater openness of the policy process. Communications and transportation technologies have made possible not only much more rapid dissemination of events as they happen, but have also made intervention in the political process more feasible. Particularly in the United States where there are so many points of leverage in the political system, this development has led to more organized public pressure on officials in a wide variety of policy areas, including foreign affairs, and a greater sense of public concern for and involvement in policy formation. The downside of the picture is that this higher level of participation has also contributed to fragmentation and blockage of the policy process. An effective leader can use these same technologies to mobilize public opinion for a particular course of action, as President Reagan was able to do for a time. The frustrations he experienced, however, growing in part out of the open policy process, also led to the incredible decision to bypass the formal process in dealing with Iran and the Contras of Nicaragua.

These developments affect different countries, even in the West, in varied ways. But it is a reasonable generalization that over time all countries will have to face a citizenry that has greater and more immediate knowledge, though not necessarily understanding, of issues that concern them, and with greater capability to make their views felt. The effects of this on relations between East and West are arguable: will publics push their leaders to accommodation, or will the sluggishness of the policy process prevent decisive action at crucial points of opportunity? Will public misunderstandings, inevitable or orchestrated, lead to pressures to avoid international agreements? Will a clever leader learn how to use these technologies to reach the public in other countries and influence policy formulation in new ways? Most likely, all possibilities will have validity to some extent and at different times. But the phenomenon of more public knowledge of and involvement in the policy process will assuredly be increasingly evident in all nations, and increasingly a factor in foreign relations.

Technology in the Policy Process

With the role of technology in East-West relations and world affairs outlined here, it is obvious that there must be extensive technological input in the formulation of foreign policy. Obvious it may be, but easy it is not. A major part of the problem is the point made repeatedly that it is never technology alone that is the issue, but the interaction of technology with the other elements of policy, including the politics of the policy process itself. Thus to represent the technological inputs adequately requires being able to mesh those factors with the many other non-technical aspects, to see their interaction, and how they affect and modify each other. That is not a task for a typical scientist or engineer, nor, for that matter, for a typical foreign policy official.

Other difficulties stem from the inaccessibility of many technological issues to those without technical training and even to many with it, the confusion over how to deal with scientific uncertainty, the danger of reliance on experts, and the difficulty of dealing with often esoteric science-based methodologies used increasingly to analyze policy issues. It is made even more difficult by the international misuse of technological information, sometimes as a mask for ideological or bureaucratic goals, and by the tendency to imagine that technology can be used as a fix to solve a political problem. The lack of an adequate mechanism for technological debate and advice in the White House these last years has demonstrated, in the case of SDI in particular, the importance of the need.

Better capacity for dealing with the technology-foreign policy interaction is needed in all governments. In the United States, it has been a perceived need for many years, with only sporadic progress in correcting the deficiencies. Presumably, other governments have similar deficiencies, and similar records in attempting to meet the need. Those deficiencies have affected East-West relations in the past, and certainly will do so in the future.

Time

There are many specific effects of technology that have altered foreign policy processes. Perhaps the most striking that deserves mention here is how technology has changed the significance and value of time. For example, the rapid-delivery characteristics of strategic weapons systems, the speed and availability of communications and transportation and the rapidity with which events become widely disseminated tend to have the effect of shortening the time available for decision-making in government. In some cases, the time is so short as to raise the question of whether a response system in fact remains under human control.

But this compression of time is accompanied by a stretching of time horizons on other issues, as the effects of current international activities of man on energy, natural resources and climate, for example, have important consequences over increasingly long time frames. Consideration of these issues must be carried out with the frames of fifty years or, in the case of radioactive waste, much longer time horizons, with inadequate models, enormous uncertainties, and heavy reliance on small communities of experts.

The combination of the two puts an important and rather new burden on the policy process. Though its effects are by no means restricted to East-West relations, the major countries of

East and West are heavily affected by them; that burden is only likely to increase in the future.

The Boundary of Foreign Affairs

The once reasonably clear separation between foreign and domestic affairs is no more. Technology has been critical to the creation of a global economy in which elements thought of in the past as domestic are now directly affected by, or themselves affect, developments in other nations. There are many implications of this change, some discussed earlier. It is worth noting, however, how important this change is to the formulation and execution of policy.

No longer are foreign policy officials necessarily dominant in the making of foreign policy. Now, the number of legitimate actors has greatly increased, and the factors that must go into a given policy are enormously expanded. In some important policy areas, foreign policy officials are often eclipsed by central bankers, trade negotiators or agriculture officials. A related change, often undervalued but quite frequently of great significance, is the schedule pressure on senior government officials arising because of the number, breadth and variety of questions with which they must deal.

Again, these are changes that do not impinge only on East-West relations, although it is certainly the case that East-West relations have tended, in the face of competition from a growing number of pressing international concerns, to maintain priority in the policy processes of the major nations. However, the agenda of policy issues continues to expand, with the interconnectedness of all becoming more evident. In that environment, resolution of East-West issues, or the search for new policy departures, will become more complex and more difficult to bring to satisfactory conclusions.

International Organizations

Almost all international organizations today have programs involving science and technology, and several were in fact created as a result of the need to deal with the consequences of technological developments. All are affected directly or indirectly by how technology impinges on their primary field of interest. And many are significantly involved in East-West relations.

However, beyond stating that there is likely to be greater institutionalization of aspects of East-West relations, this is one subject in which substantial change would not be expected in the time scale of interest. New organizations, for example to conduct cooperation in magnetic fusion, may well come about, but the organizations *per se* are unlikely to have important bearing on the climate or environment of relations. The subjects for which organizations may be created may be significant, but it is the subjects that will be significant, much more so than the organizations.

IV. Postscript

The discussion above has identified some overarching effects of technology that will condition the environment in which East-West relations take place over the coming decades. It has also insisted on the view that technology and its effects cannot be seen in isolation from their setting. Probing the effects of new technological developments can only be done by exploring their interaction with the many factors present in a particular issue or time. In fact, technological development is itself a product of a complex of factors heavily conditioned by issues that go beyond technology.

If one summary trend growing out of technology and affecting East-West relations had to be singled out, it would be that all of the relevant nations will find themselves much more deeply involved with each other, with other nations, and with issues outside the security area than is the case today. Security-related issues will remain on the agenda, while new patterns of economic and political dependency and new global issues will emerge demanding the attention of the major nations of East and West.

But in the longer run, the societal impacts of technological change may have other, more problematic, effects that may ultimately be more important than those discussed here. These effects are associated with the psychological impact of technology, which already has been a considerable, if poorly understood, force in human affairs. Attitudes of alienation and antiscience tend to emerge and retreat in social affairs, without a clear sense of their final impact. All industrialized nations report serious problems of disaffection and drugs among their youth, presumably caused at least in part by the growing impersonalization of society and by the breakdown of the close-knit family, to both of which technology has substantially contributed.

It is also evident that society is still trying to cope with the effects of the scientific revolutions that removed man from his central role in the cosmos, then reduced him to just one more cog in an impersonal evolutionary scheme, and finally gave him the power actually to destroy his entire heritage, if not his race. How important these developments are in loosening the bonds of social structure is a common focus of enquiry, but certainly they are major causes of recent social change and unrest.

Now, however, two more discoveries are in the offing, which, added to the others, will even more dramatically alter man's view of himself and his relations to his surroundings and to his fellows. They are the unravelling of the mysteries of life and of the brain. It seems to be only a matter of time, though certainly beyond the turn of the century, before both are "solved." What the effects of those achievements will be cannot be anticipated. Though they will certainly have major practical spinoffs, their most important effects will more likely be in psychological and social realms that will go well beyond the changes and problems already induced by those earlier revolutionary scientific discoveries. The question of what will be their effects on East-West relations seems almost inconsequential in the light of the much deeper and more fundamental alterations in society they are likely to bring about.

2

Technology and Social Change: East-West Comparisons

JERZY J. WIATR

Introduction

The technological revolution of the second half of this century has already created new social conditions, both in the developed capitalist countries of the West and in the socialist countries of Eastern and central Europe. While the technology itself is neutral concerning the sociopolitical division of the world, its consequences differ from society to society. The objective of this chapter is to analyze the processes of technological revolution and their impact on society in two groups of countries: the developed Western societies of Europe and North America, on the one hand, and the socialist countries of Eastern and East central Europe, on the other. I have decided to concentrate on these two groups of countries, focusing on an East-West comparison and excluding the whole variety of underdeveloped or less developed countries in the southern hemisphere.

The second industrial revolution is based on radically new technologies and scientific achievements, among which cybernetics, automation and microelectronics play the key roles. Its consequences are felt in all aspects of social life: in economic organization, social structures, culture and politics. The individual is exposed to so many fundamental changes that even a new type of man—*Homo universalis*—is anticipated.[1] Revolu-

---

1. Adam Schaff, *Wohin fuehrt der Weg? Die gesellschaftlichen Folgen der zweiten industriellen Revolution* (Wien-Munich-Zurich: Europaverlag, 1985), p. 148.

tionary changes in communications change the cognitive processes and produce a new world of *comutopia*, to borrow a term from a Japanese author.² They also lead to serious differences between world powers concerning the appropriate nature of a new information order in the space age. A Canadian political scientist compared the situation in this respect to a battlefield "of a *Blitzkrieg*, in which all the forces seem to be in motion and in which only the swift of foot can achieve their objectives," and observed that "if escape does come, it will come to those who, whether alone or in concert with others, can master the new technology and learn to bend it to social and human ends."³

A characteristic of the second industrial revolution is the overlapping of modern technologies with old technologies inherited from the past. Even more important, many contemporary societies—particularly those left behind during the first industrial revolution of the nineteenth century—have to cope with the social problems created by the new technologies when they have not yet fully mastered the consequences of the first, delayed, industrial revolution. The rapidity of economic, scientific and social changes in our time produces new problems and complicates the solution of the old ones.

Combined with this is the fact that the developed world of today is divided between different socioeconomic and political systems. The East-West division, itself the result of communist revolutions and of World War II, has two fundamental consequences: 1) it creates different institutional frameworks for technological change; and 2) it artificially divides the world into two parts, which—in spite of attempts to keep the gates of cooperation open—develop to a high degree along isolated roads.

Therefore, even though the technological challenge is a common problem for all of mankind, the character of this challenge and its social implications still differ from nation to nation.

---

2. Yoneji Masuda, *The Information Society as Post-Industrial Society* (Bethesda, MD: World Future Society, 1981).

3. John Meisel, "Communications in the Space Age: Some Canadian and International Implications," *International Political Science Review* 7, No. 3, p. 329.

Common Problems

Elsewhere I have defined four common problems stemming from the technological revolution of our time.[4] First among them is the aging of not only the old technologies but of whole sectors of the economy as well. The post-industrial society is the one in which material and human forces are being moved from the first and second sectors of the economy to the third: services and information. Changes in the class structure ensue, particularly in the composition and relative strength of the industrial working class. A new social stratum (or class) emerges, which can best be called a "new working class of the technotronic era" and which combines the elements of a new, educated working class and elements of the highly sophisticated technical intelligentsia. While the old working class still exists (in some of the most developed Western societies it is composed mainly of migrant workers from less affluent parts of the world), the emergence of the new working class complicates and alters the traditional lines of class relations and class contradictions. While the class structures of East and West differ fundamentally, they both are exposed to consequences of the technological revolution. My assertion is not that such consequences will bring a convergence of the class structures of capitalist and socialist societies, but rather that both will become fundamentally different from what they were before the arrival of the new technological era.

The second common problem is the radical transformation of the technological bases of communication. The changes in this area since World War II have no equal in human history. The use of space-based communications satellites not only decreases the distances between countries and peoples but also limits the extent to which governments can control the flow of information. They also create a new factor: the *rate* at which information can be transferred. From today's technological perspective, there is nothing to prevent two scientists, thousands of miles apart, from writing a joint scientific paper using telephone-connected computers. The moment approaches

---

4. Jerzy J. Wiatr, "Technological Challenge and Socialist Political Systems," Paper presented at a conference on "The New Technological Challenge and its Impact on Socialist Societies," organized by the United Nations University and Jagellonian University of Cracow, Mogilaby, 1987.

when not only sounds but also images will cease to have boundaries.

> The ways in which citizens perceive and respond to events and therefore the interactions between them and their governments, are bound to alter substantially. The structure and organization of private enterprises and governments will likewise undergo major changes. Furthermore, the very process of acquiring facts and values and the means through which people act will substantially alter in both the private and public spheres. The ubiquitous tension between centers and peripheries may assume entirely new forms; class distinctions will be related to the use people make of new technologies; new technocratic elites will emerge; the size and nature of communities within which individuals and groups identify themselves will reflect the capabilities of information and telematics. All these and other developments are certain to make a different political process from that known so far—both within and between states.[5]

The changes described so colorfully by John Meisel are already taking place, albeit not at the same rate, in both West and East. However, their impact on societies will differ, not so much because of different levels of technological innovation as because of the deep differences in the political arrangements within which those changes are taking place. I shall discuss this below.

The third common problem results from the basic change in the nature of war and, consequently, in relations between states, in particular between the superpowers and their alliances. In the post-Clausewitzian era, a total war is not only immoral but also totally irrational, since it cannot result in achieving any political aim.[6] The post-Clausewitzian world has been created in the laboratories of nuclear physicists. While the monster of nuclear weapons created—for the first time in human history—a real danger of a self-inflicted end to mankind, it also forced us to think in terms of a new international order in which war will no longer be the instrument of solving problems between states. A total reassessment of the theory of war became necessary, based on the assumptions that: 1) total

---

5. Meisel, "Communications in the Space Age," p. 300.

6. Jerzy J. Wiatr, "Sociology of War in the Post-Clausewitzian World," *Dialectics and Humanism* 13, No. 4 (1986), pp. 161–170.

nuclear war between the superpowers is not a rational option, and therefore cannot be rationally fought and won; 2) non-nuclear wars between the superpowers, armed with huge nuclear arsenals, are unlikely and if begun would inevitably escalate to nuclear confrontation; and 3) local wars between non-nuclear powers are possible and in fact have been fought throughout the last forty years, but their ability to serve rational political purposes is severely limited by the possibility of nuclear escalation. Because of the possibility of such escalation, even the non-nuclear wars of today are much less useful as a means to achieve political goals. Therefore, nations of the world have to learn how to solve their differences in a peaceful way. Either they will manage to learn this, or they will perish.

The fourth common problem is the contradiction between strong national aspirations—increased in particular by the access to a national cultural heritage of those social classes which previously played a secondary role, as well as by the process of decolonization of Asia and Africa—and the unifying consequences of modern technologies. Some scholars question the future of nation-states.[7] While I do not believe that we are approaching the era of the withering away of nation-states, I do believe that relations between them will have to alter fundamentally. Some forms of transnational integration will develop and they will go further than the existing forms of international integration. Taking into account the East-West division, however, it is difficult to forecast the future of such transnational integration. The fundamental common problem for both the West and the East is whether transnational integration, which is becoming inevitable in the world of new technologies, will take place only within the confines of two rival blocs of nations, or whether it will be capable of changing the nature of this division as well.

These are, to my way of thinking, the most fundamental common problems. There are other consequences of technological change that are also worthy of attention, though not of such fundamental importance. Changes in values and lifestyles, including changes in the consumption of material and non-material goods, are under way in both the affluent West and the less affluent East. An interesting phenomenon of imitation of lifestyles has taken place, contradicting to a very great extent the assumption that as the socialist societies of Eastern

---

7. Schaff, *Wohin fuehrt der Weg?*, p. 84.

Europe become more affluent their lifestyles will constitute an alternative to the old "bourgeois" lifestyle of the West. The effects of technology on the environment create common problems as well, resulting in the growing feeling that the consequences of new technological change—not exclusively beneficial—have to be dealt with as a common problem for the whole of mankind. With the progress of the second industrial revolution we will face more and more issues such as these.

The Dilemma of Retarded Development

While many of the consequences of the technological revolution are common to all countries of the world, there also exist great differences between those that achieved a high level of industrial development in the nineteenth century and those less lucky. From this perspective, differences between the developed North and the peripheral, underdeveloped South are of monumental proportions; however, they are beyond the scope of the present chapter. Albeit less acute, the problems of underdevelopment also exist in the East European socialist societies, and are creating serious obstacles along the path of the second industrial revolution.

To different degrees, all the countries of Eastern Europe—with the exception of Czechoslovakia and the German Democratic Republic—were delayed in economic development, in comparison with Western Europe and North America, in the nineteenth and early twentieth centuries. Following the socialist revolutions, however, these countries experienced a successful economic takeoff, and to a significant extent have evolved into modern industrialized societies. Their development continued the trend already set by the late pre-revolutionary regimes, but at a highly accelerated rate. W. W. Rostow, who strongly stresses the importance of Russia's developmental efforts prior to the 1917 revolution, sees the industrialization of the USSR as essentially "similar to that of Western Europe and the United States of the pre-1914 decades"—based on steel, machine tools, chemicals and electricity.[8] Overcoming underdevelopment was not the only problem the Soviet Union and the rest of Eastern Europe had to face. Historically back-

---

8. W. W. Rostow, *The Stages of Economic Growth: A Non-Communist Manifesto* (Cambridge: Cambridge University Press, 1971), p. 67.

ward in comparison with Western Europe, the whole region had to cope with the enormous consequences of the destruction wrought by World War II, and for political reasons was cut off (partly because of Stalin's distrust of the West) from closer cooperation with the technologically more advanced Western economies. Throughout the postwar years, the socialist economies of Eastern Europe kept making efforts to catch up with the West, but the results were less impressive than had been predicted. Precisely at the time when the new technological revolution was gathering strength in the West—that is, in the 1960s—the Soviet Union and the countries of Eastern Europe began to experience a slowdown of economic growth, and indeed a kind of stagnation.[9]

The Difficulties of Reform

The main problem faced by the Eastern bloc, as far as economic growth and technological innovation are concerned, is the contradiction between the system of economic organization and the demands of new technological development. The centralized system of economic management, established in the USSR in the 1930s and adopted by other socialist states after World War II, was—at best—well-suited to the demands of industrial takeoff (socialist industrialization), but unable to generate sufficient innovation and motivation for the new technological era. Its deficiencies are well-known and have been discussed by many scholars in the West.[10] In Eastern Europe, several of the best and most reform-minded economists have called for restructuring of the system at least since the mid-1950s, particularly in Poland, Hungary and Czechoslovakia. It was, however, only in Yugoslavia that—due to the rupture of political and ideological ties with the rest of the socialist bloc following the anti-Yugoslav resolution of the Cominform in 1948—fundamental changes in the economic system were introduced as early as the beginning of the

---

9. Jozef Wilczyski, "Cybernetics, Automation, and the Transition to Communism," in Carmelo Mesa-Lago and Carl Beck, eds., *Comparative Socialist Systems: Essays on Politics and Economics* (Pittsburgh: University Center for International Studies, 1975), pp. 397–417.

10. Ibid.; and Alec Nove, *The Economics of Feasible Socialism* (London: Allen and Unwin, 1983).

1950s.[11] Yugoslavia has, therefore, become a laboratory for experiments in market socialism and worker self-management from which other countries of Eastern Europe sometimes borrow ideas. However, even the Yugoslav system of a decentralized, market-oriented socialist economy has not solved all the problems stemming from historically inherited underdevelopment as well as from contradictions inherent in the process of socialist transformation. While Yugoslav achievements are impressive, there have been serious difficulties, particularly in recent years.[12]

In Poland, economic reform became a focus of much debate after the destalinization of 1956. Prominent economists such as Oskar Lange and Michael Kalecki worked out the blueprint of the economic reform, which was intended to incorporate many elements of the Yugoslav model. However, due to the unwillingness of the political leadership to accept more fundamental reforms, the Polish economy remained over-centralized and inefficient, which resulted in a slow rate of growth and recurring social tensions.[13] It is only now, after the traumatic experience of the early 1980s, that attempts are being made to reform the Polish economy in a way which—if successful—should make it better suited to the needs of radical technological innovation.

With the exception of Czechoslovakia and Hungary, East European economies have remained predominantly centralized and avoided market restructuring, at least for the time being. The Czechoslovak reform, however, was short-lived, and was all but abandoned in the post-1968 years. Only in Hungary, of all the Council for Mutual Economic Assistance (CMEA) countries, has the idea of restructuring the economy according to the needs of the market found some practical implementation. The Hungarian case has generated considerable interest abroad. While less radical than the Yugoslav one, it did produce a greater measure of autonomy for enterprises and a greater responsiveness of the economy to the rules of the

---

11. Fred W. Neal, *Titoism in Action: The Reforms in Yugoslavia after 1918* (Berkeley: University of California Press, 1958).

12. Stephen R. Sacks, *Self-Management and Efficiency: Large Corporations in Yugoslavia* (London: Allen and Unwin, 1983).

13. Janusz G. Zielinski, *Economic Reforms in Polish Industry* (New York: Oxford University Press, 1973).

market. It is not, however, a socialist market economy. Rather, in the words of a leading Hungarian reformist, it is "a mixture of the distributional consequences of both bureaucracy and market."[14] Janos Kornai stresses the fact that in Hungary—as well as in other socialist societies—there exist powerful forces which work against a radical, market-oriented economic reform, even if such a reform is necessary for economic growth, technological efficiency and innovation.[15] The present stress on economic restructuring in the USSR, as well as the growing difficulties of practically all socialist economies in Eastern Europe (with the possible exception of the GDR), may in the long run produce deeper changes. If—or when—this happens, much of the credit will have to go to the requirements of the new technological era.

Difficulties encountered by reformist economists in Eastern Europe are usually attributed to the opposition of the political bureaucracy, unwilling to give up its power over the economy as well as its privileges. While I do not intend to completely challenge this view, I am convinced that this is not the only problem. Two other dilemmas must be identified.

First, radical economic reform, based on the application of market mechanisms, implies a considerable increase of income differences and a transformation of social structure in a way that is not compatible with the tenets of Marxist-Leninist ideology. To quote Jozef Wilczynki on this subject:

> Several "birthmarks of capitalism," instead of fading away, have reappeared and have been officially embraced in the interest of economic progress. If they are eliminated in the higher phase of communism, one wonders if it will be possible to sustain high levels of productivity to enable all-round affluence, and avoid stagnation. . . . If, on the other hand, capitalist devices are retained as permanent features, then the system will not be in the image of Marxian full communism.[16]

---

14. Janos Kornai, "The Hungarian Reform Process: Visions, Hope, and Reality," *The Journal of Economic Literature* 24, No. 4 (1985), p. 1724.

15. Janos Kornai, *Growth, Shortage, Efficiency: A Macrodynamic Model of Socialist Economy* (Oxford: Oxford University Press, 1982).

16. Wilczynski, "Cybernetics, Automation and the Transition to Communism," p. 413.

It would be easy to explain away such a problem by referring to it only in terms of the ideological dogmatism of some political bureaucrats. With all its complex problems, the socialist system of the East European countries has managed to produce a system of values strongly reflecting the ideological content of Marxist ideology, at least in relation to distributional justice and the functioning of the socialist welfare state.[17] Polish empirical studies demonstrate the degree to which substantial sectors of the society—particularly the less skilled and poorly paid—remain committed to these values, and object to policies which would mean a radical shift in this respect. Moreover, a considerable part of the political elite has been socialized in these values and it would be naive to believe that it will part with them easily. In other East-bloc countries, particularly the USSR, the impact of such ideological beliefs is even stronger.

Second, not only the interests of the political bureaucracy but the interests (at least the immediate ones) of other strata constitute obstacles to radical reform. As mentioned above, those who are less skilled, older, or less adaptable to the new conditions have vested interests in maintaining the basic features of the centralized economy, even if it means a slower rate of economic growth and less affluence in the long run. Opposition to radical economic reform is being launched not only by sectors within the bureaucracy but by much broader groups in the socialist societies. In fact, the struggle for reforms accelerates interest formation and contradictions in the whole structure of socialist societies.

Because the socialist economies of Eastern Europe have been retarded in comparison with the West, the social dilemmas resulting from the needs of the technological revolution are for them more serious and deeper than for their Western rivals. At the same time, the economic and political structures of those societies, as well as their ideological values, make it more difficult for them to respond fully to the requirements of new technology. It is for this reason that the new technological revolution will produce deeper contradictions and greater tensions in the East than in the West. One aspect of such tensions is the contradictory nature of the East European working class. Its position in society and its political role are undergoing a deep

---

17. Aleksandra Jasinka and Renata Siemienska, "The Socialist Personality: A Case Study of Poland," *International Journal of Sociology* 13, No. 1 (1983), pp. 3–87.

change.[18] On the one hand, the working class suffers the most from the deficiencies of a command economy that is unable to provide sufficiently high standards of living, create safe working conditions, or to offer the most modern technological equipment and tools. On the other hand, however, it is the working class that will have to pay the price of radical economic reform, of new technologies and new economic organization—at least during the initial stages of economic reforms. Empirical studies on the impact of automation on workers in both the East and the West have demonstrated the intensity of such problems.[19] It is within the working class of Eastern Europe that the decisive struggle for the second industrial revolution will be won or lost.

Political Implications of Social Change

The question of political implications of the technological revolution has been discussed in the socialist countries of Eastern Europe for quite a long time. The emphasis has been somehow different from that of similar discussions in the West. The political consequences of new technologies have attracted less interest than the political requirements for the new technological revolution. In fact, many East European scholars have demonstrated keen interest in drawing up the blueprints for strategies of development which will make it possible to generate technological progress and simultaneously to avoid the danger of "technocracy," seen by many as incompatible with socialist democracy.

In my contribution to this debate, published many years ago,[20] I distinguished two different political strategies—"rationalized centralism" and "democratic self-government"—

---

18. Jan F. Triska and Charles Gati, eds., *Blue-Collar Workers in Eastern Europe* (London: Allen and Unwin, 1981).

19. Betty M. Jacob and Philip E. Jacob, "Motivation for Work and Technological Change: A Sixteen-Nation Comparative Study of Social and Political Consequences of Automation," Paper presented at the Ninth World Congress of the International Political Science Association, Montreal, 1973.

20. Jerzy J. Wiatr, "L'evolution des institutions politiques et la revolution scientifique et technique," *Res Publica: Revue de l'Institute Belge de Science Politique* 15, No. 1 (1973), pp. 127–128.

and expressed my preference for the second variant. This was also the point of view of many supporters of reforms in Poland, as well as in at least some other socialist countries.

I have not changed my general perspective. Technological progress without democracy is dangerous, since it will inevitably lead to technocratic deformations. Moreover, I seriously doubt that without democracy—that is, without the involvement of large social groups, workers and technical intelligentsia in particular—it is possible to break the barriers built by conservative vested interests. Therefore it is necessary to adopt the second of my two alternative strategies.

From the perspective of fifteen years I can only stress the fact that although I saw the difficulties and contradictions of the process of technological revolution, the reality—in Poland particularly—significantly surpassed my predictions. The contradictions have proved to be deeper and more dangerous in their consequences than I previously believed and their character has proved much more complex, extending beyond the technocratic threat. In particular, the experiences of Poland in the 1980s gives one a lot to think about.

In Poland, as in the majority of socialist countries, the necessity to reform the economic system became evident in the mid-1960s. Economic development slowed down, and if a high rate of growth was maintained in some countries it took place because of high foreign borrowing or drastically limited consumption. The rationality of the unreformed, centralized system was coming to an end.

Therefore it was only natural that the second half of the 1960s and the first years of the following decade witnessed serious social and political upheavals in several socialist countries. These upheavals had many unique characteristic features. None was a simple copy of another. National preconditions and specific characteristics of the various political systems were reflected in each. By and large, however, they constituted a clear sign that socialist countries had come to a historical turning point. Qualitative changes in economic and political systems became necessary.

Necessary, but not inevitable, at least for the time being. Only in Yugoslavia, following the historic plenary session of the Central Committee at Brioni in 1966, and in Hungary, following the adoption of the New Economic Mechanism in 1968, were the reformist trends strong and lasting. In Poland they have been prevented from consolidating themselves, and in Czechoslovakia they were long ago prevented from achieving their goals. In the other European socialist states pressures for reforms have not materialized. The most important historical fact was that the Soviet Union—after a period of quite stormy, unprepared and not very skillfully realized reforms from 1953 through 1964—entered a long period of stabilization, criticized today as one of stagnation. The conservative opposition to reforms in the USSR was capable not only of postponing badly needed changes in the Soviet Union but also of influencing most of the other countries of the region in the same way. Postponing the reforms resulted in a situation in which reforms are now even more dramatically necessary than twenty years earlier.

Nowhere did the delay of the reforms bring more dramatic consequences than in Poland. The attempt to substitute structural reforms with economic growth artificially stimulated by loans from Western banks led to a collapse of the economy and the emergence of strong social pressures for change. Between the summer of 1980 and the end of 1981, Poland's socialist political system was on the brink of collapse and the nation was threatened by civil war. To divert such danger the extraordinary measure of martial law was invoked, but the need for deep reforms was not eliminated in this way. Quite the contrary. It was precisely after 1981 that economic and political reforms were undertaken in Poland with a scope larger than in any previous period of Poland's postwar history. At the same time, these reforms are often, and not without cause, criticized as not going far enough and not being quite consistently realized.

In a different, much safer way, the Soviet Union has also arrived at the threshold of systemic transformations. The policy of "restructuring" adopted by the 27th Congress of the CPSU calls for qualitative reforms in economic and political spheres. Since pressure from below had not been particularly strong (in fact, the passivity of the Soviet population is often identified as a serious obstacle to rapid change), this strategy was adopted because of the realization by the elite of the need to change important elements of the system. Poland and the Soviet Union, although united in their drive for reforms, represent two different situations. The change in Poland originated from a strong pressure from below and is now put into effect from above, in the context of a society actively pressing for change. In the USSR, change comes from above and has yet to generate more active support from below. Other European socialist states, with the exception of Yugoslavia and Hungary, have yet to decide which way to go. Unless they adopt the strategy of reform from above—and put it into effect fast enough—they may find themselves in a situation not very different from the Polish experience of the early 1980s. There is no way the socialist countries can avoid economic and political reforms. It is only the strategy of reforms, their speed and depth, and consequently their long-term effects that remain uncertain at this stage of the game.

II
Technology, Arms and Disarmament

3

The Impact of Technology on Nuclear Deterrence and Strategic Arms Control

JOSEPH S. NYE, JR.

I. The Problem of Volatility

We live in a world of volatile technology. Nuclear technology has transformed the potential destructiveness of warfare. Very high-speed integrated circuits allow us to master unprecedented amounts of information as well as miniaturize weapons. Infra-red sensors and adaptive optics in the visible spectrum allow us to detect things not possible a decade ago. Robotics allow operations under previously impossible conditions. New materials will permit superconductivity to be applied in a variety of contexts. Genetic engineering may open new potential for destruction that we have not even begun to understand. But perhaps the most important question for the next decade or two will be whether technology will begin to reverse the dominance of offense over defense that has thus far characterized the nuclear era.

In the early 1970s there was an apparent U.S. consensus on strategic stability that involved (1) maintaining survivable U.S. forces; (2) forgoing the capability to destroy a large part of the Soviet arsenal; and (3) the idea that defense was not yet technically ripe for deployment.[1] Some have argued that this consen-

---

I am grateful to Albert Carnesale and Ashton B. Carter for helpful suggestions.

1. Paul Stockton, *Strategic Stability Between the Superpowers* (London: International Institute for Strategic Studies *Adelphi Paper* 213, 1986), p. 13.

sus was always illusory. Others argue that it rested too heavily on the assumption that vulnerability is desirable and inescapable.[2] Such critics argue that the Soviet Union never fully accepted such a concept of stability with its narrow technological focus.

Nevertheless, the U.S. and USSR tried to codify the strategic balance of the 1970s. The idea of parity and a balance between offense and defense were critical assumptions of the SALT agreements of that period. By the late 1970s, however, even sympathetic observers were concerned about the problems that technology posed for these efforts to codify the strategic balance through arms control. Christoph Bertram identified three technological challenges: first, the speed of technological change was injecting a high degree of ambiguity into restrictions directed primarily at quantitative levels of forces; second, qualitative improvements were often more significant but less verifiable than quantities of weapons; and third, the trend of technological change was towards multi-mission weapons which undermined the definitional categories which had been a primary organizing principle.[3]

In the early 1980s, some hoped to cope with these effects of volatile technology by instituting a nuclear weapons freeze. Attractive though the idea might appear at first glance, it had several problems. It was difficult to conceive of ways to freeze both measures and their countermeasures in a verifiable fashion. For example, it was easier to freeze construction of new submarines or bombers than to freeze the development of anti-submarine warfare or air defense technology. The problem was complicated by the fact that many of the technological developments with greatest significance for weapons systems, such as data processing and new materials, had large areas of overlap with civilian technology. Obviously, civil development could not be frozen. Furthermore, even if a freeze on technology could be implemented in a symmetrical way, it might prevent the development of some systems which in retrospect could have had a stabilizing effect. For example, if there had been a freeze in 1959, it might have prevented the deployment of bal-

---

2. Colin Gray, *Nuclear Strategy and National Style* (Boston: Hamilton Press, 1986), ch. 5.

3. Christoph Bertram, "Arms Control and Technological Change," in UNESCO, *Arms Control and Disarmament*, 1981, p. 150.

listic missile submarines, which proved to have a beneficial effect on strategic stability because of their relative invulnerability.

Perhaps we do not need to worry so much about this dilemma of technological volatility. For example, Samuel Huntington has suggested that qualitative arms races have rarely led to war in the past.[4] Others might argue that minor changes in the nuclear balance are relatively unimportant. In their view, the nuclear balance is enormously robust because of the unprecedented destructiveness of even a single nuclear weapon.[5] While there may be truth in this observation, it is also worth noting that statesmen have tended to be far more concerned about marginal changes in strategic balances that this observation would imply. Politics does not follow technology in any simple sense. If there were major changes in the relationship of offense and defense, the effects could be quite dramatic politically, even if they were based on imperfect technology. As Eugene Skolnikoff has put it in his overview of "Technological Factors Shaping East-West Relations," "Identifying technological trends and exploring them one by one is unlikely to be successful because it belies the interaction of technology and politics . . . . All important technological issues in international politics must ultimately be dealt with in political terms."[6]

II. Stable Deterrence

The stability of deterrence is an elusive concept with at least three dimensions: crisis stability, arms race stability and political stability. Each dimension is affected by technological developments, although technology alone does not determine stability. American strategists have tended to stress technical determinants of stability, while Soviet strategists tend to pay more attention to political factors. As Table 1 illustrates, both technical and political factors are important.

---

4. Samuel P. Huntington, "Arms Races: Prerequisites and Results," *Public Policy*, 1958, pp. 41–82.

5. Robert Jervis, *The Illogic of American Nuclear Strategy* (Ithaca: Cornell University Press, 1984).

6. See Eugene Skolnikoff's chapter in this volume.

Table 1
Determinants of Stable Deterrence

	Technical	Political
A. Crisis Stability	Invulnerable second strike capability and C^3I	Crisis prevention and management
B. Arms Race Stability	Slow technological development/deployment	Confidence-building and arms control measures
C. Political Stability	First-strike threat and denial capability	Agreements, rules and regimes

Crisis stability refers to the absence of incentive to strike preemptively in times of crisis. The key is the survivability of significant forces after suffering a first strike so that effective retaliation is assured. Much depends on the vulnerability of weapons. For example, the vulnerability of ICBMs depends upon the technological race between accuracy and detection, on the one hand, versus hardening and mobility, on the other. However, vulnerability is a matter of degree and varies with the number of weapons as well as with technological progress. If both sides were reduced to only a few weapons, in the desperate circumstances when a nuclear war appeared likely the temptation to believe in the possibility of limiting damage by a preemptive strike would be increased. With many weapons, the degree of invulnerability need not be so great in order to discourage any such temptation. The lower the number of weapons, the higher the premium on the invulnerability of those remaining and the greater the perturbations caused by unforeseen technological changes.

Political factors also affect crisis stability. Indeed, Soviet commentators sometimes describe the American concern with crisis stability as too technical and mechanistic. It is not merely the characteristics of the weapons systems but also the ability of the parties to communicate and understand communications. Tacit and written rules of prudence for the prevention and management of crises may be as important as the technical characteristics of the forces. Nonetheless, if one looks at Soviet investments in the hardening and mobility of land-based forces and investment in ballistic missile submarines, it is clear that they are concerned about the technical as well the political dimension of crisis stability.

Arms race stability refers to the absence of incentive for either side to rapidly expand or modernize its arsenal. It depends in part on technological opportunities and in part on reactions to the programs of the other side. Thus, the phrase "action-reaction." Since strategic nuclear systems often involve lead times of more than a decade, arms race stability depends on anticipation of what the other side's technological development program may be. Worst-case appraisals of how the other side might react to a given technological opportunity often affect decisions before evidence is fully available. The fear of "breakout" from a treaty or rapid change in armament levels can foster arms race instability.

Arms race stability is affected by the number of weapons as well as by the technological opportunities. The prospect of achieving a significant advantage from rapid changes in technology is less likely at high levels of weaponry. If the strategic arsenals were at very low levels, the prospect of achieving significant advantage through rapid change could appear more attractive. Unless there were a high degree of political cooperation and confidence, arms race stability would then be more difficult to maintain.

This indicates that political factors are also very important in determining arms race stability. Advocates of arms control argue that the very process of reaching agreement on reductions or limitations generates confidence and cooperation. In this sense, arms control can be seen as confidence-building which limits the worst-case assumptions which otherwise accelerate budgets for technological development. Moreover, to the extent that arms control treaties embody provisions enhancing transparency and communication (such as not interfering with national technical means of detection or providing accurate numbers to the other side), they limit worst-case analyses and thus reduce arms race instability.

Political stability refers to the effectiveness of deterrence in reducing incentives for major coercive political changes. It is sometimes forgotten that deterrence is not merely about avoiding nuclear war, but also about avoiding major losses of critical foreign policy stakes. Unless one includes that political dimension of stable deterrence, one cannot understand the failure of both sides to simply maximize crisis stability. In fact, there are trade-offs between the crisis stability and political stability dimensions of deterrence. The desire to avoid coercive political change and loss of foreign policy stakes leads both sides to invest in technological developments and forces which allow them to threaten the other side as well as to deny the other side a prospect for victory. Table 2 illustrates the contradictions

Table 2
Alternative Force Structures

Preferred Forces for Political Stability		Preferred Forces for Crisis Stability	
"US"	"THEM"	"US"	"THEM"
1. invulnerable	vulnerable	1. invulnerable	invulnerable
2. invulnerable	invulnerable	2. invulnerable	vulnerable
3. vulnerable	vulnerable	3. vulnerable	vulnerable
4. vulnerable	invulnerable	4. vulnerable	invulnerable

between the technical characteristics of the forces desired for crisis stability and those desired for political stability. In terms of crisis stability, it is best for both sides to have invulnerable forces. In terms of political stability, which requires a capacity to threaten, each side has a strategic doctrine to pose a credible threat to the forces of the other side. Each side continues technological developments to enhance such capacity.

When Americans sometimes charge that the large Soviet ICBM force in fixed silos is destabilizing in its technical characteristics, they are clearly referring to crisis stability. Soviet strategists often reply by saying that such a force is stabilizing because it frightens the United States and thus discourages Washington from attempting political coercion or other adventures that might lead to a crisis and war. Obviously the Soviets are referring to political stability in such a debate. Some American strategists believe that a similar capability on their part is necessary for the credibility of the American foreign policy of extended deterrence. For example, proponents of the MX missile have argued that being able to threaten Soviet missile silos or at least having an equal capacity to promptly threaten hard targets is important to the deterrence of Soviet actions that could otherwise lead to war. In other words, Americans use an analogous argument to that used by the Soviets in terms of relating the technology of the forces to the purpose of deterring political coercion and protecting foreign policy stakes.

Political stability rests on more than such technical characteristics. It is also affected by perceptions of relative numbers of weapons as well as broader geopolitical factors. The degree of cooperation and understanding surrounding the military balance in particular regions is an important factor. Geographical asymmetries are also important. For example, the Soviet Union is ringed by states with modest nuclear arsenals while the U.S. is not. Conversely, the United States points to the Soviet

Union's advantage in conventional forces and its geographical contiguity to the areas crucial to world politics. The U.S. argues that extended deterrence in those areas depends upon a credible prospect of nuclear use, not on the mere equality of numbers. Efforts to create credible prospects of use have often led to an increase in the number of weapons or in types of weapons—witness the original arguments for deployment of INF as well as the arguments for MX.

It is hard to argue, however, that extended deterrence and political stability in crucial areas such as Europe are most strongly affected by technological factors related to strategic nuclear weapons. On the contrary, more important may be the relative stakes the two parties have in the area and their relative conventional force capabilities. In that sense technology that improves the theater force balance can indirectly affect the strategic balance. One of the remarkable points about extended deterrence is that it has lasted long after its critics proclaimed its demise—witness General DeGaulle's statements a quarter-century ago. In part, this may reflect the critics' overconcentration on the strategic nuclear factor. It may be that the high stakes and negative burden of resolve are the most important factors in the political function of what is known as extended deterrence. Moreover, conventional and theater nuclear capabilities which prevent either side from believing that it could achieve a quick *fait accompli* and therefore present a residual risk of escalation create an enormous uncertainty that discourages temptations to exploit geopolitical advantages. In short, the political determinants of political stability are probably more important than the technological determinants.

With all this said, however, deterrence still depends upon some prospect of nuclear use, and extended deterrence is more difficult to make credible than deterrence of attacks against one's homeland. How much prospective use and what type is a matter of considerable controversy. There are two ways nuclear use could occur—deliberately or inadvertently. Consequently, there are two forms of deterrence. One relies upon credible threats of deliberate use; the other relies upon the chance of inadvertent use or what has been called the "threat from chance." The two types can be labelled deliberate and inherent deterrence.[7] Those who believe that inherent deterrence is suf-

---

7. See Joseph S. Nye, Jr., *Nuclear Ethics* (New York: The Free Press, 1986), pp. 107 ff.

ficient can be relatively relaxed about technological developments which seem to believers in deliberate deterrence to threaten the strategic balance. Those who believe that inherent deterrence is sufficient are willing to accept very deep cuts without great concern. Just so long as there are enough weapons to pose a threat to cities and civilization, and so long as there are not effective defenses to prevent such a threat, there will be deterrence. Because of the destructiveness of nuclear weapons, even a small and accidental prospect of use goes a long way in providing deterrence.

On the other hand, those who believe that some degree of deliberate deterrence is necessary have to be concerned about the way in which reductions or technological developments would affect the ability to threaten particular military targets. Those who wish to maximize deliberate deterrence will press technological developments which may be bad for survivability and crisis stability, but that will increase the credibility of deliberate use. From this perspective, the idea of non-nuclear strategic weapons appears attractive. The technology of guidance and accuracy may make such systems feasible in the next decade, but the political uncertainties surrounding them are far from being resolved.[8] Would the victim of a non-nuclear strategic attack ride out the loss of his silos rather than launch his nuclear weapons? No one can be sure such a war would remain limited.

In practice, there is a continuum between those two types of deterrence and many strategies fall somewhere between them, rather than at the two ends of the spectrum. If one accepts a doctrine of assured destruction, one could settle for a few hundred large invulnerable weapons and target only cities. But as a deliberate strategy, the destruction of cities lacks credibility, appears genocidal, and could be suicidal. On the other hand, a strategy of prompt attack against silos may increase rather than limit damage if it leads the side being attacked to a policy of launching on warning those missiles that might otherwise have been withheld. The question of how much and what type of counterforce capabilities are necessary for stable deterrence underlies differences among strategists not only

---

8. Carl Builder, *The Prospects and Implications of Non-nuclear Means for Strategic Conflict* (London: International Institute for Strategic Studies *Adelphi Paper* 200, 1985).

about numbers of weapons but also about the desirability of technological changes that increase counterforce capabilities.

A third viewpoint would argue that the more appropriate targets for deterrence are neither cities nor silos but the military forces of the two sides—in other words, those forces that would carry out invasions or military operations. Threatening such forces denies them their military objective as well as punishes the other side while still leaving cities unharmed. The cities' continued existence provides incentive to negotiate a termination of any war. But such a third alternative doctrine of "countercombatant targeting" requires technological developments in both real time target acquisition and in retargeting capabilities to be effective.[9] Nonetheless, technological developments do seem to be moving in this direction. Of course, how one evaluates these technological changes depends in large part on how one approaches the issue of what is necessary for stable nuclear deterrence.

III. Strategic Defense

Thus far this discussion has assumed a world in which the technology of the offense will continue to prevail over the defense in the strategic realm, as it has since the beginning of the nuclear age. But technological changes that greatly enhance the capability of defensive systems might change the nature of nuclear deterrence. Before the nuclear era, deterrence involved aspects of both punishment and denial. In the nuclear age thus far, it has been the fear of devastating punishment which has been the major deterrent. In the pre-nuclear era, fear of being unable to accomplish one's objectives also served as a deterrent. The advocates of defense dominance in the nuclear era do not necessarily wish to replace deterrence in the broad sense as much as to shift the mix of deterring elements from heavy reliance on punishment to a greater reliance on denial. Table 3 sketches a set of alternative strategies ranging from those such as assured destruction, which rest almost entirely on punishment, to those such as defense dominance which rest almost entirely on denial.[10]

---

9. See Nye, fn. 7, for discussion.

10. This table draws upon but differs from Colin Gray, cited, ch. 9.

Table 3
Alternative Strategies

	Technical Requirements	Political Problems
Punishment		
1. Assured Destruction	A few hundred survivable weapons	Is it credible for extended deterrence?
2. AD + limited options	#1 + a few hundred fairly accurate weapons and fair C³I	Will both sides cooperate in limiting war?
3. AD + countervailing	#1 + many prompt hard target weapons against silos, leadership and C³I targets. Survivable C³I.	Will limited attacks on cities be distinguishable from MAD? Can control be maintained?
4. AD + countercombatant targeting	#1 + survivable, fairly accurate, low yield weapons. Improved C³I and target acquisition.	Can collateral damage be limited? Will both sides cooperate in limits?
5. AD + partial defense	#1 + partial active and passive defenses (BMD, Air, ASW, Civil).	Does it increase crisis instability and arms race instability?
6. Defense Dominance	Near perfect survivable defenses (BMD, Air, Civil).	Is it stable without extensive political cooperation? Is the transition manageable?
Denial		

Obviously no strategy is perfect and all have political problems associated with them. The technological requirements are simplest for those at the punishment end of the spectrum. The major impact of technology on strategic deterrence and arms control in the next decade or two will be those which make strategies 5 and 6 (i.e., partial defense or defense dominance) feasible options. That is the issue to which I now turn, while noting that advances in the technology of offensive weapons also present possibilities for increasing the element of denial and reducing the element of punishment in deterrence strategy (e.g., option 4).

The ABM Treaty was signed in the early 1970s when it became clear to both sides that protecting cities with missile defenses was not within the reach of then existing technologies. Americans viewed it as contributing both to the arms race and to crisis stability. If the best available defensive systems at the time could be defeated by additional offensive missiles at a fraction of the cost, then actual ABM deployments would simply drive the other side to acquire more offense: the result would be a wasteful offensive/defensive "spiral" of new weapons with no net increase in security for either side. Similarly, it was argued that limited missile defenses would be more useful to the side striking first, because the defense would stand a far better chance of handling a ragged response than a full-scale first strike. And, of course, there would be a fear that the other side might preempt for precisely the same reason.

The ABM Treaty is sometimes characterized as a commitment to accept mutual vulnerability as a desirable state of affairs for all time, but this claims too much for the treaty. Vulnerability would hardly be desirable if there were an alternative concept that promised both reduced exposure and a lesser risk of instability. Nor is BMD inherently incompatible with our strategy of second-strike deterrence. Missile defenses that are unambiguously limited to protecting our land-based retaliatory forces might in some circumstances improve deterrence by increasing attack uncertainty or raising the price of attack to a level greatly in excess of the value of the targets. Even then, however, any decision to deploy point defenses would have to be based on a net assessment that the extra protection in deterrent terms was worth the cost of allowing the other side to go forward with its own defenses. In the early 1970s, both sides decided they would be better off if neither deployed extensive missile defenses (including hard point defenses) than if both sides did.

While President Reagan's 1983 speech renewed the public debate over that conclusion, the objectives of his SDI programs remain complex and confusing. On one hand, the president's goal has been clearly nothing less than a defense-dominant world in which entire populations are invulnerable to attacks by nuclear missiles. As former Secretary of Defense Caspar Weinberger said: "There is only one SDI . . . when the President says that we are aiming at a strategic defense designed to protect people, that is exactly what he means."[11]

Conversely, the kinds of missile defense that other SDI supporters favor would be aimed at frustrating Soviet attacks

---

11. Remarks by former Secretary of Defense Caspar Weinberger at the John F. Kennedy School of Government, Harvard University, September 5, 1986.

against military forces. One alternative would be point defense of hardened missile sites and command structure with short-range (possibly nuclear-tipped) rockets that would intercept incoming warheads within a few miles of their targets. Another would include a more elaborate exoatmospheric system to break up highly structured attacks, increasing the attacker's uncertainty and limiting the extent of damage inflicted on both military targets and populated areas. In both cases, however, the aim would be to strengthen deterrence, based on the assured punishment of a second strike as well as denying the opponent any temptation to strike first. There is a basic tension in the underlying policy objectives of the Strategic Defense Initiative. The president's stated objective was to replace deterrence based on punishment, whereas a number of proponents justify it in terms of enhancing the current type of deterrence. The difference matters because the optimal technologies for the two objectives are not always the same. Thus far, however, the program has managed to blur such choices.

It is still too early to say how the trends in technology will affect the prospects for either version of strategic defense. There has been significant technological evolution, but the trends do not indicate at this stage what defense objectives might be plausible over the long term.[12]

Surveillance, Acquisition, Tracking and Kill Assessment (SATKA)

The technologies for detecting the launch of Soviet ballistic missiles are already well established in present generation U.S. early-warning satellites that stare at Soviet missile deployment areas from geosynchronous orbit. For support of active defenses, improvements would be required to pinpoint individual boosters and to track them. Even more challenging problems are tracking warheads and distinguishing them from decoys. Historically, mid-course discrimination has been a great barrier to advances in BMD. Discrimination and surveillance with passive means remain highly problematic because the signals emitted from warheads and decoys can be manipulated to alter their appearance; active means (i.e., using radar

---

12. This section is based on the conclusions of the Aspen Strategy Group, *The Strategic Defense Initiative and American Security* (Lanham, MD: University Press of America, 1987).

images) seem equally unattractive at this stage, although work continues.

"Interactive" discrimination by impact from a low-power neutral particle beam or laser may allow calculation of the mass of an object. Among the problems, however, are the survivability of the beam generator, the survivability, size and structure of the sensor system necessary to collect the radiation returns from the target object, the problem of separating signals from the "background noise" of nuclear explosions, and how to handle the enormous amount of signals collected in the process. Generally the same sorts of problems apply to "kill assessment."

Directed Energy Weapons (DEWs)

Lasers and particle beams often receive the most public attention. In theory, DEWs would be useful as boost-phase weapons, given the extremely short engagement times available to the defense and the greater vulnerability of boosters (relative to harder warheads) to the thermal and other effects of beam weapons. They are also being considered for mid-course applications. Even so, progress thus far has been modest— steady but not dramatic. Because of their vulnerability, space-basing of the weapons themselves is regarded less favorably than it was a few years ago. Moreover, chemical lasers now appear unlikely to satisfy the requisite brightness and survivability criteria that would make them effective devices for boost-phase intercept.

Experiments in 1985 suggested that free-electron lasers (FELs) might be very efficient in converting power to usable energy. Current research is investigating whether FELs could produce beams of appropriate wavelengths and with similar efficiencies at power levels approaching SDI requirements (many orders of magnitude greater than those achieved to date). If FELs proved feasible as weapons, they would most likely be based at installations in parts of the country where cloud cover and atmospheric distortion of the beam would be minimal, such as on mountaintops or in desert areas. As ground-based systems, they would require complex optical systems in order to fire their beams up through the atmosphere and space-based mirrors to relay and refocus their beams onto the targets. Work on such "adaptive" optics and space-based mirrors has picked up accordingly.

A more uncertain but potentially higher payoff concept is the X-ray laser driven by a nuclear explosion in space. Because X-rays do not propagate into the atmosphere, the system would have to be space-based or "popped up" prior to use. The unknowns surrounding X-ray lasers are enormous. What has been demonstrated to date is that lasing can be achieved; the capacities of such a weapon to convert energy efficiently, to focus the beam and to point the beam are not yet understood.

Kinetic Energy Weapons (KEWs)

Recent efforts to make early deployment possible have led the SDI program to increase its research and development on kinetic energy weapons concepts. Although the speed of attack advantage lies with beam weapons, KEWs might offer greater kill capability against hard targets (e.g., warheads) and a greater diversity of deployment modes. KEW work aims to develop very high velocity projectiles of such accuracy that they could deliver small explosive charges to within a few meters of the targets or even strike them directly. For boost- or post-boost-phase defense, SDI is working on a space-based kinetic energy kill vehicle (SBKKV) system that would fire a 5–7 kilogram warhead at velocities approaching 7 kilometers per second. Very low warhead weight is required if the space-based interceptor rockets are to be kept light enough to deploy in space affordably. Electromagnetic rail-guns that deliver a payload at twice the speed are also being explored but (in view of their power requirements and homing problems) not apparently as a near-term prospect. In addition, ground-based rockets for late mid-course or high endoatmospheric intercept are being developed under the Exoatmospheric Reentry-vehicle Interceptor Subsystem (ERIS) and the High Endoatmospheric Defense Interceptor (HEDI) programs; and a program called FLAGE aims to develop a small, very maneuverable rocket for intercepting warheads in the terminal phase of attack.

Current R&D in the KEW area addresses interceptor weight, guidance and propulsion. Advances in focal plane sensor technology are being exploited as a way to move toward the low-mass guidance specifications for the SBKKV concept. Infra-red terminal homing and impact kill have been demonstrated to some extent in the highly publicized Homing Overlay Experiment of 1984, in which an interceptor projectile successfully collided with a missile warhead in mid-course. Nonetheless, the jump from technology development to more integrated components and KEW systems will require major new innovations in areas such as highly maneuverable homing warheads, battle management, and (over the long term) breakthroughs in affordable space basing, a solution to the mid-course discrimination problem, and a solution to the ASAT problem.

Supporting Technologies and Battle Management

Perhaps the most difficult challenges to SDI over the long term lie not with the weapons or sensors, but in the support functions of space lift, power systems, data processing, and overall battle management. Technologies in these areas will be critical in determining the cost-effectiveness of various weapons or SATKA concepts, how much of the overall system can be space-based, and whether each layer of the defense can be managed coherently as part of a complex system which can be operated at high levels of readiness on extremely short notice.

The dimensions of the power-supply challenge for space systems are fairly well understood at this point, though solutions seem very distant. SDI is pursuing a program to develop a space-qualified nuclear reactor that could be scaled up to meet the power levels required for peacetime "housekeeping" of sensors or weapons; initial space testing of a prototype might begin in the mid-1990s. Work on the multi-megawatt power sources needed for actual battle engagement is only in the conceptual stages.

The question of launch capacity is affected by uncertainties about what proportion of the overall system would be based in space. It does seem, however, that a launch system with many times the lift efficiency of the current shuttle will be needed. SDIO is now evaluating design concepts for economical space launch systems with the aim of driving down the current real launch costs of about $3,000 per pound to levels approaching a few hundred dollars per pound.

The command and control issues surrounding SDI are even less well understood and more controversial. The operational reliability of large-scale space-based defenses would be a major challenge. Unless the system were tested under the stresses of nuclear explosions in space, our understanding of its combat effectiveness would always be extremely uncertain. In addition, it is not really known at this stage whether it will be possible to develop the kind of software necessary to manage reliably the enormous flow of data among the various components (i.e., sensors, command, weapons) of each layer and between layers within the overall system. Faced with a major attack, the task of assuring central management of battle functions, including tracking all objects in space, would probably be overwhelming unless the system could "home in" on the warheads and ignore the decoys. Avenues for progress may be found in distributed data-processing architectures, and ones in which computational requirements are defined by fault-tolerance rather than error-free criteria for all but the highest risk functions. But the data processing dimension may still prove to be the Achilles heel of the system.

It is too soon to draw clear conclusions about whether these technological trends will present attractive options for a change of strategy in the 1990s. Forecasting the net effect of technology is risky. Major breakthroughs could change the way we look at problems. Conversely, research often raises new challenges even while providing answers to old questions.

Thus far, the technological progress has been incremental. Innovations have occurred across a broad range of technologies and at a pace generally consistent with increased funding, but there have not been dramatic breakthroughs. Looking ahead, more basic research on FEL and X-ray laser technology may yield by 1993 conceptual and possibly experimental advances. In areas where physical principles are better understood, such as adaptive optics, beam relay and "smart" kinetic energy rockets, continued innovations will probably bring these technologies close to weapons criteria by the early 1990s. Progress in chemical lasers and many SATKA programs may be slower unless conceptual or design innovations occur.

None of these developments, however, would reshape the fundamental challenges. The Reagan administration set for itself two important criteria by which to judge the desirability of any potential defensive system; first, it must be capable of surviving direct attack, and second, it must be cost-effective at the margin (i.e., cheaper to augment than to overcome with additional offensive weapons). At this time, there is little reason to accept the views of those who claim that a survivable and cost-effective space-based system will be ready for a deployment decision now or in the early 1990s. Even the prospect of developing technologies for cost-effective layered missile defense before the end of the century remains forbidding.

Viewing the issue strictly in technological terms, the gap

between where we are now and large-scale operational systems is enormous, especially in the areas of survivability, directed energy systems and support functions. Many years of work lie ahead. But perhaps the more significant point is that while we should expect technological innovation, we cannot be certain that it will yield the sought-after solution.

For one thing, the same technology development that aids the defense can also help the offense. Many of the systems we are now investigating would vastly improve the capacity of the attacking side to saturate or suppress the defense. Economical space lift, for instance, would dramatically change the character of the threat. Small terminal-homing rockets and, in the longer term, the X-ray laser offer certain advantages to the offensive side, which, by definition, has the luxury of choosing the place and time to launch an attack. Moreover, effective defense in the boost phase seems problematic in the face of a responsive offense. The structure of boost-phase intercept is such that a space-based defense may simply be too exposed to sustain itself amid a variety of technically feasible countermeasures, including the use of nuclear weapons to blind or destroy the components. "Pop-up" systems might be less vulnerable but extremely difficult to redeploy within a useful time frame. Finally, defense in the terminal phase is also problematic in view of the current non-nuclear terms of reference for the SDI program. In the terminal phase, nuclear weapons could work to the defender's advantage, but SDI's insistence on non-nuclear KEWs may give an insurmountable bonus to an attacker armed with the new technology of maneuvering warheads.

These observations apply to the horizon of foreseeable technological change during the next decade. Beyond this, the path is uncertain.[13] One critical area in which a technological breakthrough could make a major difference in changing the present offense/defense balance is in the mid-course region. There, the timing and modes of intercept might favor the defending side, if only a means could be found to discriminate reliably between warheads and decoys. Until that point, however, and perhaps considerably beyond it, we are not likely to witness a long-term trend in high technologies leading us away from an offense-dominant world.

---

13. Harold Brown, "Is SDI Technically Feasible?" *Foreign Affairs* 64, No. 3, 1986.

Even if these technological projections prove wrong, and there are major breakthroughs which enhance the advantage of the defense over the offense, major political problems will remain in moving from the current strategies based on punishment to strategies emphasizing defense and denial. Imagine that perfect defense is possible. One frequently noted problem is the transition process. How will the two parties react if one appears to be approaching a goal of perfect defense more rapidly than the other? Will this lead to political crises or changes in the risk aversion that the parties show in the management of crises? Unless there is a high degree of political cooperation, proposals for sharing technology to assure symmetry at the point of arrival at perfect defense seem highly implausible, particularly since many of the technologies such as data processing and new sensors are the same technologies which will make an enormous difference to the military balance on the conventional battlefield.

Even if the transition problem is managed adequately, the conditions for stability under an imagined perfect defense remain a puzzle. As Charles Glaser has claimed, a world of perfect defense would share some of the same problems of stability as a world of zero nuclear weapons.[14] In a world with high levels of offensive weaponry, small technological perturbations or increases in numbers do not make much difference to overall stability. But in a world of zero nuclear weapons, the hiding or reinvention of even a few weapons would make a large difference. Similarly, in a world in which each side had what it believed to be perfect defenses, marginal changes in the technology of defense penetration by one side or the other could have quite large perturbing effects on the balance. In other words, smaller technological changes may have larger political effects in a world where each side had what it believed to be perfect defenses than in the current world.

One might try to get around this problem by assuming that technology will remain static and that improvements in either offenses or defenses will be impossible, but such an assumption would be foolish in the extreme. It is for these reasons that a number of people have observed that any prospect that defense might enhance strategic stability must depend very heavily not merely on technology but on a large degree of

---

14. Charles Glaser, "Why Even Good Defenses May Be Bad," *International Security* 9, No. 2 (Fall 1984).

political cooperation. For example, in a world in which the challenges for the defense were not to outwit a responsive opponent but to protect against cheating or advances by new entrants into the nuclear game, defenses might play a more manageable role. The concept of SDI as an insurance policy, however, is a far cry from the type of programs now being pursued. In summary, moving to a strategy of defense/comprehensive denial (#6) will require heroic changes in technology and in political cooperation.

Even a movement to a strategy of partial defenses (#5) will require major technological and political assumptions. The problem is not our technological ability to add defenses; we can do that today. The problem is to add defenses in a way which enhances the three dimensions of stable deterrence. It is sometimes said that the addition of defenses complicates a first strike. This may be true, but it also complicates a second strike (unless one engages in the fallacy of the last move, i.e., assuming that one's own side gets defenses but the other side does not). In such a world, one has to ask whether the means of complicating a first strike by adding defense is cheaper than other means of complicating a first strike such as investing in the technologies related to mobility, hardening and concealment. Not only must partial defense be cost-effective in comparison to the alternatives, but it must be also cost-effective in comparison to the prospects for increased penetration which technology is making possible for the offense. Otherwise the efforts to add limited defense will merely precipitate an offense-defense race to the disadvantage of the defense. In short, limited defenses raise prospects of both crisis instability and arms race instability unless one assumes a cost-effective technology and/or a high degree of political cooperation. Arms control agreements that regulate the relationship between the offense and defense, or which constrain the degree of the offensive threat may help. Once again the assessment of the impact of technology on stable deterrence cannot be taken on the basis of technological trends alone but must include a political context.

IV. The Role of Arms Control

As noted at the beginning of this chapter, the volatility of technology presents serious problems for efforts at arms control. Technology cannot be frozen and efforts to limit one class of weapons sometimes turn out to be like punching a pillow. As one area is compressed, a bulge comes out in another place. Moreover, arms control has often been oversold by its proponents. The public has been lead to believe that reducing numbers of weapons is similar to reducing risks of nuclear war when that is not necessarily the case. For example, the shift from bombers to missiles led to a major reduction in weapons (e.g., from bombs to missile warheads) in the early 1960s, but it also shortened response time from eight hours to thirty minutes. Reduction in numbers is not the same as reducing nuclear risks. Even reduction in the number of destabilizing weapons (assuming one can reach agreement on definitions) could change the structure of military forces but not affect the operation of those forces. It is the command, control and operation of nuclear forces during crises which is most clearly related to the probability of nuclear war. Moreover, political steps to prevent and manage crises may do more than reductions to improve the U.S.-Soviet relationship and thereby lessen risks.

Contrary to conventional wisdom, unless one assumes major political changes, nuclear reductions may not save money. They might save money on strategic weapons—which only account for 15 to 20 percent of the defense budgets of the two countries—but could increase overall defense costs by putting more pressure on the conventional forces. Nor would reducing the number of nuclear weapons necessarily reduce the odds of accidental use. Here number is not the major factor behind the odds of accidental war. The quality of technical devices such as electronic combination locks, redundancy in warning systems and development of special systems to ensure reliable command, control and communications are far more important than the number of weapons.

A recent study of the arms control record of the past few decades came to a number of interesting conclusions.[15] Progress in arms control has taken place only when neither side has had an advantage. Meaningful constraints in any particular category of weapons were achieved only when neither side really wanted the weapons or programs constrained. The arms control process and arms control agreements tended to codify existing defense plans. Nevertheless, the arms control process and arms control agreements did reduce uncertainties in pro-

---

15. Albert Carnesale, "Learning from Experience with Arms Control" (Cambridge, MA: Kennedy School of Government mimeo, 1986).

jections of each other's forces. Contrary to the claims of some critics, the arms control process and agreements neither lulled the United States into spending less nor stimulated it to spend more than it should on defense. In short, the conclusions of the report suggest that the major benefits of arms control were in the political relationship and in the reduction of uncertainty. In other words, despite the enormous uncertainties and problems created by volatile technology, arms control still played a useful (if modest) role.

If one is to avoid the debilitating impact of technology on arms control, however, one has to conceive of arms control less as a set of legal agreements which freeze a technology or military balance and more as a process of communication about the management of a dynamic balance. The arms control process has contributed to learning about the realities of the nuclear balance and the difficulty of maintaining nuclear stability on both sides.[16] Reduction of uncertainty is important since it limits the tendency toward worst-case analyses which would otherwise take place. In fact, one can argue that the main function of arms control is to increase transparency and communication in the management of the military balance between the two countries.

This is not to belittle the importance of formal agreements or efforts to reach reductions. Informal operational arms control and formal negotiated reductions need not be opposed alternatives. They can be complementary. In fact, the most important aspects of the SALT treaties were the provisions on open skies for satellite reconnaissance, agreed counting rules for various types of weapons and the Standing Consultative Commission.

If arms control is viewed as a way of managing a dynamic balance rather than trying to freeze a static balance, then it remains an essential part of the response to changing technology, as the discussion above has indicated, in a number of particular instances. Not all arms control will be formal nor need all arms control deal with reductions. In some cases, freezes and bans on testing may slow certain technologies, but one should not expect technology to stand still. The critical questions are to find ways to discuss changes in technology and the way they affect the different concepts of stability if both sides are to reach their objective of avoiding nuclear war while main-

---

16. Joseph S. Nye, Jr., "Nuclear Learning and U.S.-Soviet Security Regimes," *International Organization* 41 (Summer 1987).

taining important foreign policy goals. Arms control should provide occasion for continued communication at various levels of government relating to managing the strategic balance. Such communications can include both high political officials, high military officials, and technical experts in bodies such as the Standing Consultative Commission. From this perspective, volatile technology need not destroy the prospects for arms control. On the contrary, it should increase the demand for it.

4

Advanced Technology and European Security: Conceptual Considerations

ANDRZEJ KARKOSZKA

Introduction

There are several reasons why conventional weapons and forces associated with them should be a focus of European arms limitation efforts in the coming years. First, the success of the Stockholm conference on confidence- and security-building measures created a positive environment for further steps. The momentum created by such steps could bring about far-reaching restraints and limitations on armed forces in Europe, thus consolidating the CSCE process and leading to a peaceful and stable Europe.

Second, the newly signed U.S.-Soviet agreement on intermediate-range nuclear forces offers promise for a significant decrease in the number of nuclear weapons in Europe. Though the remaining nuclear stocks are still massive, the prospect of even partial reductions has caused alarm in some West European and U.S. political circles about the diminishing military security of the West, resulting from the alleged imbalance between WTO and NATO conventional forces. The military situation in Europe consists of two elements, nuclear and conventional. As a rule, a number of states have been reluctant to embark on limitations in only one of these two fields, arguing that such an approach undermines security in the other. This has resulted in stagnation in both areas. Therefore it is more appropriate to confront jointly, at the negotiating table, the avowed security concerns of both sides in the conventional field.

Third, the constant evolution of conventional weapons towards more potent and efficient means of warfare is now entering its mature stage. In all categories, very destructive, very efficient, complex weapons systems have been developed, even replacing tactical nuclear weapons in some operational roles. The importance of these weapons in the European security equation is growing rapidly.

Fourth, seen in the light of the ongoing debate on prospective military postures, the available data on the actual level of technology allows one to assume that several new categories of conventional weapons systems will appear on European soil in the near future. These include: long-range reconnaissance-strike systems usually connected with the NATO concept of Follow-on Forces Attack (FOFA); stealth aircraft and missiles; conventionally armed ballistic missiles of tactical, short and intermediate ranges; and tactical ballistic defense systems. These systems, which will soon be deployed in Europe, will represent a dramatic increase in the offensive potential of European conventional forces. Because they are so powerful and decisive for the outcome of a potential conflict, these systems are preemption-prone in a serious political crisis. Thus, their appearance will be one of the most destabilizing developments in Europe. The prevention of their deployment would be beneficial for the security of the continent and may save substantial resources for all states concerned. Failure to prevent the new deployments will force European states to regain their sense of security only by an extended arms race and heavy emphasis on military spending and production. Furthermore, the relaxed political atmosphere that has been gradually emerging on an international level will suffer. Its decline will be exacerbated by adversarial politics in other spheres of international relations and by hostile propaganda used to justify and sustain the military effort, a task made easier because the new weapons systems will be more menacing and more difficult to assess soberly. Their deployment will surely make any progress in arms limitation in Europe doubtful; even if some agreements were to be signed, their meaning would be limited.

In the long run, a new round of the arms race in Europe, exemplified by the deployment of new conventional weapons, may deepen the political division between East and West. This would not only slow down progress in the security domain but also eliminate any prospect for further civilian cooperation in a wide spectrum of economic, trade, technological and scientific fields. Yet this occurs at a time when closer cooperation is indispensable in all aspects of life in view of the increased interdependence of states, especially of the European states.

It is vital that we recognize the possible consequences of the new round of the arms race looming ahead and act to prevent it before it begins. We now confront both greater-than-ever dangers and costs of an unrestricted drive for better weapons and greater-than-ever opportunities for improved political relations between East and West.

However, it will not be enough to call for the resumption of arms limitation talks in the traditional sense. The same strategies will not suffice. Urgent, new and more imaginative action is needed—first, because the traditional approach to arms control and disarmament is under heavy criticism as not being responsive to constantly changing military technology and technological change generally; second, because conventional weapons are a more complex matter than nuclear weapons, embracing a much broader technological area and remaining closely related to non-military technology; and third, because technology in the general sense of the term, encompassing both civilian and military technology, is undergoing a rapid transformation and assuming a growing importance in international relations.

Before offering any alternatives or contributions to the existing framework of East-West negotiations (that will be undertaken in the final part of this chapter), I will thoroughly present the argument for conventional arms limitations. This requires a presentation of the actual state of conventional weapons technology and, subsequently, a cursory analysis of the benefits and dangers connected with the preservation of European security through the unrestricted expansion of military potentials, including the expansion of conventional weapons technology. Furthermore, it is important to examine military and civilian technology as a dominant new element of the European security process.

I. The Need for Conventional Arms Limitations

A New Era of Conventional Arms

The history of military technological development indicates several shifts in the relative importance of various areas of technology, as exemplified by particular weapons systems. After more than three decades of focus on nuclear weapons technology and the delivery systems associated with them, and after the deployment of thousands of these systems, the pace of their expansion has subsided. There is no need for more of them and not too much room for any greater sophistication. The third generation of nuclear weapons, with the sole exception of neutron weapons, is still to be developed and is associated more with space applications than with ground- or sea-based forces. Over the last decade it is conventional weaponry that has undergone the most profound technological development. The time for its deployment has now arrived and will last until the beginning of the next decade. The era of space weapons is soon to dawn upon us, although concrete results remain far in the future. Technologically, however, the two military areas—advanced conventional weapons and space weapons—have more in common than meets the eye. If not controlled or restrained, they will feed on each other's achievements.

The dimensions of the qualitative change in conventional arsenals is difficult to grasp in full. Some consider this process a revolutionary one, although in fact we should speak of a gradual, evolutionary process which went on for many years largely unnoticed and has only recently caused alarm. It is the multitude of small incremental changes which brought about entirely new classes of weapons, the consequences of which are comparable to those caused in the past by such truly revolutionary developments as gunpowder, steam power, the machine gun, the tank, and—during recent decades—nuclear power, intercontinental delivery vehicles and satellites. Such developments were considered revolutionary; they dramatically altered military tactics and doctrines, the organization of military forces, the relative position of different states, and international relations as a whole. Modern conventional weapons may be just as profound in their implications.

The new generation of conventional weapons—some recently deployed, others soon to be deployed—possesses operational capabilities unknown a decade ago. These weapons have the ability to strike targets with high accuracy at any distance, as long as the targets have been located and identified by auxiliary reconnaissance systems. They are independent of weather and time of day. They can attack a target with nearly no time delay after its discovery. They have enormous speed and freedom to maneuver and fire, in addition to the mobility of weapons and forces themselves. Conventional weapons can be tailored to all types of targets—point and hard, large and soft, fixed and mobile, surface and underground. This last capability makes them candidates for the execution of roles so far ascribed only to nuclear weapons. Conventional weapons function with greater reliability and efficiency and can interact freely on the battlefield in a synergistic way, mutually enhancing their effects. Conventional firepower, measured by the amount of ammunition delivered in a given time over a given area, has increased by several orders of magnitude. Conventional arms tend to bypass traditional environmental classifications of different weapons systems: Sea-based weapons may be projected on land; air forces are now more than ever decisive for naval and land operations; and soon some land-based weapons may become dual-capable, also operating in the air and in space.

The general picture of new conventional weapons technology presented above depicts a much more potent category of arms than ever before. It fails, however, to convey the sense of instability that some of these conventional weapons systems have brought to the military situation in Europe.

Among the most destabilizing weapons systems, now in an advanced stage of development, are the long-range reconnaissance and strike systems, in which sensors will be able to locate targets deep in the opponent's territory and to vector various long-range delivery vehicles onto the targets. The delivery vehicles may be ground- or air-based, the latter with standoff capabilities. Exemplifying such systems are the so-called TABAS system, based on a space-capable Saturn missile; the T-16 missile, itself a derivation of the Patriot missile; the T-22 missile, based on Lance; the modification of Pershing II called CAM-40; the new ballistic missile called ATACMS; the long-range standoff cruise missile; and various remotely piloted vehicles or drones. The MLRS unguided rocket launcher is a good example of one of the weapons already deployed which, after some modification, will be used for the same purposes.

These weapons will be able to deliver a great variety of munitions and submunitions, the effects of which will be tailored to meet operational needs. They will be able to destroy targets that are stationary or mobile, multiple or single, and exposed or well protected. Even a single salvo from a submunition dispenser will cover a vast area with precise destructive power. This large family of weapons includes: the MW-1 dispenser; air-delivered munitions such as the Durandal and BAP-100, the AST 1228 and the JP 233; and the STABO, Skeet, KB-44 and MIFF submunitions.

All of the weapons mentioned under the category of reconnaissance and strike weapons systems will be connected by a multitude of communication links, permitting jam-resistant command and control. This C^{3}I network will consist of systems such as the Joint Surveillance and Target Attack Radar System (JSTARS), the Joint Tactical Fusion System, the Joint Tactical Information Distribution System, and AWACS backed by satellite platforms, such as Milstar and NAVSTAR/GPS.[1]

Another potent new category of weapons soon to be reckoned with in the European theater is that of stealth aircraft and cruise missiles. Their impact will come mainly in their capability to deliver an attack with greater impunity from air-defense systems and with greatly diminished warning time for the defenses of the opponent. This type of weapon will thus be best suited for a surprise or short-warning operation.

Electronic warfare systems will be an indispensable element of new conventional weapons deployment. Their widespread use will confuse the whole defense structure and render the overall military situation more unpredictable.

With nuclear weapons playing a diminished role in the future European theater, the possibility exists that several categories of today's nuclear-armed ballistic missiles may be equipped with conventional warheads. These missiles could well be used as a prelude to the main aircraft offensive, crippling anti-air defenses and thus exposing a potential enemy to massive and unopposed destruction.

The prospect of numerous conventional ballistic missiles calls for a necessary counterweapon in the form of anti-tactical ballistic missile defense (ATBM) systems, first clustered around defense assets and later spreading into a nationwide ATBM network. If both sides develop the ballistic offensive and defensive systems in a parallel fashion, the situation may be considered relatively stable. However, such parallelism may not be obtainable. If one of the two sides develops and deploys one of the two categories earlier than the other—or, even worse, if both of them race to do so—the ensuing military situation may be of the utmost danger.

The weapons discussed above are currently in various stages of development. Some are already being deployed; others are

---

1. On this subject see in particular: *SIPRI Yearbook 1984*, pp. 281–318, and *SIPRI Yearbook 1986*, pp. 193–208. (London: Taylor and Francis, 1984 and 1986).

in final testing or, like the potential ATBM systems, will be available in five to ten years. All are part of weapons systems of the immediate future that will have a profound impact, direct and indirect, on the European military and political scene.

The primary influence of these weapons systems stems from their dramatically increased offensive capabilities over the conventional weapons currently deployed in Europe. Capable of a surprise attack, they are at the same time so powerful that in case of serious crisis their destruction becomes a matter of first priority for a side threatened by them. This development would tend to reduce the time available to political leaders and military commanders for a sober assessment of the situation and of the true intentions of an opponent in crisis. Without a doubt, in a time of stable political relations between adversaries, the existence of any weapons need not be regarded as a direct and immediate threat. However, the expansion of preemption-prone military capabilities does not bode well for future political relations and will be a reminder of potential hostile intent and a permanent source of mistrust.

Moreover, all of the systems soon to be deployed—in particular, the ATBM systems—are or would be extremely costly. From today's perspective many of the weapons in question are beyond the fiscal capabilities of most of the individual European states contemplating their deployment. Nevertheless, the resources of these states may be pooled and the process of deployment may be gradual. What is important is the fact that once the decision on deployment is reached, reversal is particularly difficult to achieve and all negative consequences will follow.

New conventional weapons provide the best proof for the argument that in the confrontation of modern and massive armies in Europe, it is not only numbers but, increasingly, qualities of weapons, patterns of force deployment, levels of force readiness, and the general preparedness of all necessary military support structures which matter in maintaining stability. While numbers of forces and weapons play a role as an operational indicator of military potential, they matter even more as reference points in a political debate or public discussion. They are decisive when the numerical discrepancy between opposing forces is strikingly large and the military equilibrium is held at a substantially lower level (at which a potential attrition rate would cripple a technologically superior force in a battle against a more numerous enemy with less-advanced technology).

These points about numerical balances, however, have little, and certainly not decisive, relevance to the present European situation. First, the conventional balance is now—and probably will remain for the foreseeable future—under the shadow of nuclear weapons, even in the event of substantial reductions in nuclear arsenals. Second, the levels of forces are very high and are characterized by a wide panoply of types of weapons. Discrepancies, even by a wide margin, arrived at by static "bean counting" on paper do not make a difference in an operationally perceptible way. There is no objective way to envisage a theoretical outcome of hostilities between these numbers, furnished with vastly differing operational characteristics, taking into account all support systems, geography, level of training and morale of soldiers, and other factors. Even without nuclear weapons, no responsible leader could venture an aggression in Europe that would have any degree of certainty of success. Even without nuclear weapons, a conventional deterrence in Europe does exist and is enhanced by the obvious lack of political or economic incentive for an aggressive policy. The term "conventional deterrence" has a meaning quite similar to the term "nuclear deterrence," despite the vastly different physical nature of the two kinds of weapons concerned. The deterrent value of modern conventional military postures comes from the enormously expanded destructive potential of these weapons and forces, as well as from the fact that in the case of war in a highly urbanized Europe, in which the territories of both the attacked and the attacker would surely be exposed to massive air and missile strikes, the population of either side could not sustain its existence for long.

Although we tend to think in terms of numbers, this approach to military security in Europe is outdated. Politicians and publics alike, if informed and educated objectively, would end their fixation on numbers and would instead turn their attention to the real problems of the present, and especially the future balance of conventional forces—their technological expansion, their destructiveness, their destabilizing features and their growing relevance to the general arms race.

Benefits and Dangers of Military Technology

The progress of science and technology is considered nearly tantamount to the advancement of civilization. It is inconceivable to halt it altogether or reverse it. Thus the technology for nuclear weapons, or for any other modern weapons cannot be disinvented and erased from human memory. More importantly, it is often, and to some extent rightly, pointed out that several achievements of military technology have had profound positive effects on international security and stability. Such has been the case with reconnaissance satellites, which influence strategic stability by offering a solution to the verification conundrum for the two great powers operating them. One could also argue that the constant modernization of nuclear weapons has made them safer for storage and operation and has diminished the prospect of the total annihilation of civilian populations through miniaturization of nuclear warheads and the possibility of surgical counterforce strikes.

This type of argument is unconvincing, however, for the simple reason that the positive effect of that type of technology is more of a side effect. The primary effect is much more menacing. In the case of reconnaissance satellites, it is their role to provide the basis for strategic targeting, warfighting and general war preparation. In the case of the increasing sophistication of nuclear weapons and the concomitant elaboration of more complex nuclear warfighting scenarios, the question of what is considered more beneficial for the international community is highly subjective and problematic. Be that as it may, the idea of a positive effect from the steady expansion of military technology has support in various circles of Western political establishments, mainly due to their self-serving belief that it is superior Western technology that keeps the "adventurous and expansionist East" at bay. The great problem with this belief is that it is difficult to disprove in an objective manner; it is entrenched in Western propaganda, and is easily argued through a short-term and segmented approach to military developments. Moreover, such beliefs, like all kinds of beliefs, are ideologically and culturally motivated and thus resistant to alteration.

The validity of a proposition on the utility of military technology in preserving and even augmenting a state's security finds its supreme corollary in nuclear deterrence theory. An additional and equally unchallenged rationale is invoked in justifying the relentless fixation with technological solutions to the problem of the conventional balance in Europe. We are often told that superior Western military technology is the best and most secure way of compensating for the East's allegedly superior numbers. This assertion is used to justify all weapons procurement programs, no matter how excessive, and is often rationalized by producing more or less theoretical scenarios, some of which may even (who knows?) contain a good deal of truth. There is, however, one constant and troublesome feature of this attitude: it requires a ceaseless effort to make nuclear deterrence and the technologically determined balance credible. This means continual technical improvement constantly racing against the permanently demanding odds of the worst-case scenarios, since the greatest credibility is perceived in possession of either qualitative superiority or a lead in a given category of weapons. This approach, by definition, precludes common development of general and applied technology by East and West and propels division and militarized competition.

It is not my intent here to examine the very concepts of deterrence, nuclear and conventional, and determine their validity. Indeed, such an analysis would be largely futile. The actual impact of military technology, exemplified in subsequent generations of nuclear and conventional weapons systems, on military relations between East and West could be indisputably verified only in an actual war. Short of this kind of final verification, all statements are theoretically allowed, although some appear to be more plausible than others, depending on subjective perceptions and biases. What can be easily observed over the last forty years of qualitative—and quantitative—competition is the temporary character of any technical advantage. Even when it has seemed that one side's quality could compensate for the numbers of its opponent, the credibility of such an assumption has always been precarious. The whole political and military context has been too dynamic and fluid to enable states to rest assured for any length of time. No temporary advantage could be preserved for long; nor has the balance of quality versus quantity been satisfactory and unquestioned for all political forces and states concerned.

The Necessity of Conventional Arms Limitation in Europe

Throughout the postwar period, the preservation of European security has been excessively burdensome and detrimental to the political and economic cooperation of the European states. Within this context, should Europe be looking for any dramatic departure from the traditional, parallel, incremental expansion of respective military potentials, which at present is fueled chiefly by technological improvements? Some experts think not. After all, it is undeniable that both the allied and neutral European states, and in particular the two great nuclear powers, which are inextricably enmeshed in European security issues, have sufficient economic and technical resources to sustain this expansion and have adequate experience in the management of their relationship to avoid an outbreak of war, barring irrational or accidental conflict. Why then should they embark on a different track, as yet unproven and thus uncertain, of limiting military potentials through restraints on their technological competition?

In fact, there are several good reasons for a reversal of this well-entrenched policy. First and foremost, despite the postwar division of the European continent and existing ideological and political controversies between the NATO and WTO states, all these states, as well as the neutral and nonaligned ones, are engaged in a positive and evolutionary process of mending their relationships. This process, which started with the Helsinki Final Act, is still fragile. It can be fortified only by concrete and far-reaching steps in the military domain, reaching deeper and further than the present agreed-upon CBM measures. Today's Europe lives in an environment much more conducive to a major rearrangement of security relationships than ever before. There are no more territorial claims, no officially questioned borders, no open irredentism.

Second, the military competition between NATO and the WTO is visibly fueled mainly by a military-technological drive, ostensibly for the benign purpose of preserving the stability of the situation without an open goal of gaining superiority or preparing for aggression. This assertion is corroborated by a number of official pronouncements on both sides. Military technology, however, may threaten this relatively benign attitude, at least in the eyes of the prospective opponent. Thus new military technology, introduced by one side in the name of stability, eventually contradicts this justification the moment the other side introduces similar counter-technology. The vicious repetition of this cycle perpetuates basic mistrust and fortifies the heavy reliance of European states on military guarantees of security.

It is often observed that one of the sources of mistrust between the two alliances in Europe is their declaratory and operational doctrine, be it readiness to use nuclear weapons in the early stages of a conflict or readiness to thwart any aggression and carry hostilities onto enemy territory. In the prospective preemption-prone military equilibrium, these doctrines may easily be perceived as even more threatening. But even today these doctrines are founded not only on geography, historical experiences and scientific analysis—each of which influences NATO and WTO thinking in a particular way—but also on the material realities of their military forces, including technology. Their present technologies and operational doctrines are intertwined. If not interrupted by a rational decision, this will continue in the future, as doctrines adapt to advances in military technology. It is impossible to alter military doctrine without altering its material basis, whether quantitative or qualitative. It is impossible to exert much pressure on decision-makers to institute any material limitations on a state's military security if the doctrines of the opponent are not changed so as to lessen the perception of threat. New military technology, by virtue of being less known and less tested, complicates threat assessments. It tends to reinforce current military doctrines, adding an aura of expanding efficiency and effectiveness and thus fortifying their negative impact on East-West relations in Europe.

It goes without saying that the new military technologies are bound to be more expensive, that they will require more R&D resources, more inter-alliance collaboration, and more stringent control over East-West civilian technological cooperation. The whole economic aspect of European security will, therefore, be increasingly influenced in a direction which flies in the face of the objective necessity to expand common efforts across the dividing line to protect the environment, exchange information and people, expand trade, improve health care, explore new energy sources and better utilize scarce resources.

The new conventional weapons systems will cost a degree of magnitude more than their predecessors. Apart from the U.S. and Soviet economies, no other national economy will be able to sustain this burden alone. Military collaboration in both the East and West will enter into a qualitatively new stage. Entire branches of science and technology, as well as whole industries, will be mobilized and organized into the effort. In such an environment there will be less and less room for wider all-European cooperation. It may well be that the final result of the progression of military technology will be a world of space-based ballistic defenses, impenetrable air and ground-based ballistic missile networks, laser-fortified strongholds at the borders, perfect anti-submarine barriers, and so on. No offensive action would be conceivable in such an ideal military world. There are, however, two disturbing questions which must be answered now: first, are these wonder weapons affordable and workable; and, second, how are we to get from here to there? Some lesser issues include concerns about the outlook for the European continent, how national economies will be influenced, and considerations about the social and political processes connected with this drive for militarily perfect security. It is thus questionable whether this new method of preserving East-West security, exclusively concerned with rearrangements and limitations on offensive military developments, is any better than the traditional approach.

There are several important reasons to pay more attention to conventional weapons and forces in examining the problem of military technology and its influence on European security. The urge to turn more attention toward the conventional aspects of the military competition comes from a growing awareness of their influence on military security in Europe and from their above-stated technological unity with nuclear weapons. We are accustomed to emphasizing nuclear aspects of security in Europe. These weapons undoubtedly pose a more immediate, existential threat and have traditionally raised the attention of public opinion and politicians. Over the past several decades, conventional weapons were considered merely a secondary issue. This perspective is no longer tenable. Apart from the technological linkage between nuclear and conventional arms, fortified with every new generation of weapons, there is another, more conspicuous linkage: the operational relationship between the two on the potential battlefield—a close relationship of which European states are very much aware.

The political importance of this perception is particularly apparent at those times when international disarmament negotiations are centered on only one of these weapons categories. When the prospect of a far-reaching agreement is in sight, strong doubts are raised about the wisdom of such a step, which is said to jeopardize the delicate balance of security based on the existing mix of nuclear and conventional forces in Europe. The latest wave of criticism of this kind (largely West European and to a lesser degree American), which arose in the post-Reykjavik period and was invigorated by the INF agreement between the Soviet Union and the United States, proves the point. While some of this criticism stems from a general phobia toward arms control and any form of accommodation with the WTO countries, the argument about the functional linkage of conventional and nuclear weapons in the preservation of European security cannot be taken lightly. This linkage has been pointed to on several occasions during successive periods of European dialogue on military security issues, from the debate on the Rapacki Plan in 1958–1962 to the Vienna Mutual Force Reduction (MFR) talks and the discussions on the Palme Commission's nuclear-free corridor proposal. It is interesting to point out that the linkage has been invoked by both sides, although on different occasions: the West did so during the debate over the denuclearization of central Europe, the East in the Vienna MFR talks, and the West again in criticizing the corridor idea. Now, when the partial denuclearization of Europe seems to be imminent—thanks to the U.S.-Soviet INF pact—the issue of conventional weapons and forces becomes salient again and is being raised forcefully by West Europeans. The linkage should not be construed, however, as a demand for a simultaneous negotiation on measures applied to both ends of the military spectrum. This would immediately cripple the momentum of the process that began with INF. The two sides of the military security coin have to be treated in parallel; they should be conscientiously interrelated in the political debate and, if possible, during the negotiating process. We should not allow a chance to lower the nuclear arsenals in Europe to slip away due to the disregard of negotiators for the legitimate security concerns of a number of European states.

Another aspect of the argument justifying the need for conventional arms limitation measures in Europe is the fact that these types of weapons are much more directly related to the CSCE process than are nuclear arms. If the military aspects of this process are to help sustain its continuation and are not to stand in the way of progress in other "baskets" of the CSCE process, then it will not be sufficient to arrange for partial transparency, greater predictability and some token restrictions on peacetime military activities. More profound steps are called for. It could be argued that what is needed in the quest for conventional arms limitations in Europe is a breakthrough in MFR-type negotiations, permitting a swift reduction of numerical strengths. Such a breakthrough and subsequent agreement would undeniably have a substantial positive impact on the political atmosphere in Europe. In some parts of

Europe, including the WTO countries, it would be considered a major improvement in the continent's security. However, such a measure might not be considered positive by some NATO countries and, in fact, would not address the more fundamental security issues looming in the not-too-distant European future.

Today we must expand the framework of arms limitation negotiations and turn, on an international basis, toward future problems, instead of reacting to them after they are already well entrenched. Not only should weapons and forces already in existence be subject to negotiations, but future military developments as well. It is obvious, however, that any enlargement of the existing machinery of arms limitation negotiations in Europe must meet at least two difficult criteria: political usefulness for all parties concerned and practicality in other than purely political terms. With the unexpected and profound evolution of the socialist states in all disarmament negotiations, several long-standing beliefs and unquestioned tenets came under critical scrutiny. The reluctance of several Western states to enter the unknown territory of serious arms limitations was exposed; apparently they preferred the existing military status quo to the unpredictable results of such agreements.

Apart from these legitimate concerns about future security, there is an even larger question looming before us: how will far-reaching agreements be compatible with the underlying ideological and political contradictions between East and West? The East answered this question by moving from a position of peaceful coexistence to that of interdependence and by launching bold disarmament initiatives.

Notwithstanding these fundamental questions, there are other political reasons why progress in arms limitations in Europe is not welcomed by some Western nations. These reasons are connected with the lack of meaningful and perceptible gains to be achieved by those national institutions which will bear the burdens of the reversal of the arms race and the limitation of military potentials: the military, heavy industry, scientific and engineering institutions and trade companies. This is especially relevant to limitations imposed on conventional arms. Because of their technological scope, such limitations would generate much more sweeping changes in internal and international economic and trade relations than would be the case with nuclear weapons. Thus, any negotiations on far-reaching measures designed not only to cut numbers of weapons but to prevent further expansion of conventional weapons technology must have a clear understanding of these more complex interconnections.

II. Technology — An Indispensable Element of European Security

The International Impact of Technology

Among the dynamic factors influencing the relative position of states and, more generally, the actual state of international relations (e.g., demographics, economics, cultural development and social change), technology is fast becoming the most potent.[2] The scientific-technological revolution (STR), barely evident at the beginning of this century, began in earnest after World War II. Having begun with accelerated progress in the areas of machine technology, electronics, and automation, the STR is visibly moving towards still uncertain applications of biotechnology, vacuum materials technology, smart robots, artificial intelligence and other advanced industrial technological innovations. The pace of this process is steadily increasing, surpassing the human capacity for adaptation in terms of human biology, ecology, social structures and—of special concern here—international behavior.

Modern science and technology not only altered the former relationships among states but also entered into the substance of these relations. As Eugene B. Skolnikoff sees it, the global implications of science and technology consist of a steady diminution of a nation's freedom of action to apply science and technology as it sees fit, at times even within its own borders.[3]

This effect is paralleled by a need for international means to operate and regulate global technology, and by a growing gap

---

2. See Anne G. Keatley, ed., *Technological Frontiers and Foreign Relations* (Washington, DC: National Academy Press, 1985); and Stephen M. Shaffer and Lisa Robock Shaffer, *The Politics of International Cooperation: A Comparison of U.S. Experience in Space and Security* (Denver, Colorado: Monograph Series in World Affairs, Graduate School of International Studies, 1980).

3. Eugene B. Skolnikoff, *Science, Technology, and American Foreign Policy* (Cambridge: The MIT Press, 1967), pp. 3-4.

between the requirements of technology and the size and resources of nation-states.[4] Though omnipresent and all-encompassing, the consequences of various technological applications are difficult to ascertain until actually experienced. Even then they are difficult to fully comprehend because of the synergistic impact of diverse social, economic, political and environmental factors. There are numerous historical examples that show that the social and political consequences of new technology usually elude prediction.

Because they are so unpredictable, advances in science and technology will always be a potentially destabilizing factor in international relations, depending heavily on the way they are applied. That is why "one of the major concerns in the development of a powerful new technology is the possibility of its uncontrolled use for narrow national ends."[5]

A driving force behind the speed of technological progress after World War II was the fact that technology embodied one of the most visible aspects of political competition between East and West. Dictated by the West, which was certain of its superiority in this domain, technological warfare pervades all aspects of East-West relations and is still taken as the main means of pursuing the security and economic interests of the West vis-a-vis the socialist states.[6] As one Western analysis has put it:

> The pace of technological change is set by the rivalry between the United States and the Soviet Union: because that rivalry involves European allies, the Western European countries have had little choice but to adopt the most advanced military technology for their own forces as well. The technological arms race has thus required heavy expenditures on military research and development. . . .[7]

---

4. Ibid., p. 303.

5. Ibid., pp. 308-310.

6. Stephen T. Possony and J.E. Pournelle, *The Strategy of Technology, Winning the Decisive War* (Cambridge: University Press of Cambridge, 1970).

7. Judith Reppy and Philip Gummet, "Economic and Technological Issues in the NATO Alliance," in Catherine McArdle Kelleher and Gale A. Mattox, eds., *Evolving European Defense Policies* (Lexington: Lexington Books, 1987), p. 22.

The European states are taking part in the overall technological competition and they aspire, at a minimum, not to lag behind in the race. However, not all of them have the necessary objective conditions to keep apace. The technology-induced stratification of European states defies ideological and political divisions, on the one hand, and the cultural and economic unity of the continent, on the other. In the first case, the picture is somewhat blurred by the fact that the European socialist states are believed to be trailing behind the West European states technologically. In many technological fields, however, the situation is reversed, especially if the Soviet Union is taken into account. Moreover, among the West European states technological diversification is clearly visible and may grow, despite efforts to the contrary. It would be foolhardy to argue that technology is always divisive. It can also induce unity and even uniformity. Technology binds states together with expanding networks of communication, computer links and rail grids; it forces states to react in a commonly organized way to ecological decay; and its requirements urge states to share in ever more costly R&D efforts.

One could say that humanity is preordained by many of the aforementioned dynamic factors (including technology) to come ever closer together, if not actually to unite—unless, of course, we use our technology to remove ourselves from the planet. Such grand cliches notwithstanding, the process of the adaptation of international relations to new challenges—technological, economic and so forth—is unpredictable. The process may be fast or slow, or it may be partially reversed; it may cause tensions and conflicts, leaving some states ahead of others; it may be more or less profitable for various states and their respective societies. The point here is that in spite of the objective unifying tendencies of modern technology, there exist undeniable technological factors that restrain interstate cooperation and, in the case of ideological and political opponents, engender suspicion and fear. In the civilian domain, such is the case with some of the most expensive and economically most profitable technologies, which are often guarded by those states possessing them in order to secure their profits and position. This is always the case with military technology.

For Skolnikoff, the restriction on international cooperation in science and technology is natural:

> It is that science is too closely bound to the central objective of government—that of providing for the defense of the nation—

for a country ever to let control over its scientific development be shared with other nations *unless* a prior decision has been made to integrate defense responsibilities substantially with those other nations.[8]

Many analysts stress, however, that even between states sharing security interests, as among NATO members, technological cooperation and trade in high-technology items is characterized by a "growing tendency towards economic nationalism."[9] These writers see trends towards technological protectionism and restrictions on imports from and exports to other countries as unproductive, since technology is increasingly transferable across borders, and as politically harmful, as it may open rifts between some of the most industrialized Western countries.[10]

In pursuing technological warfare against the East, the United States and, more reluctantly, its allies in Europe and in other regions are constantly trying to tighten the flow of technology to the socialist states. To this end, the U.S. Department of Defense has increased the number of items on the Militarily Critical Technologies List and strengthened cooperation with countries of COCOM (the Coordinating Committee for East-West Trade Policy) and other states on a bilateral basis.[11] This effort is supported by an argument about the importance of Western technology for the defense production of the East and the impossibility of differentiating between "dual-use" technologies and those which can only be used for civilian purposes. Nevertheless, all these efforts, though perceived by many as justified on security grounds, seem doubtful for sev-

---

8. Skolnikoff, *Science, Technology and American Foreign Policy*, p. 183.

9. See Harold B. Malmgren, "Technological Challenges to National Economic Policies of the West," *The Washington Quarterly* 10, No. 2 (Spring 1987), p. 21; and Peter Maass, "EC's Complaint: High-Tech Bonn Won't Share," *International Herald Tribune*, March 31, 1987.

10. Malmgren, "Technological Challenges," pp. 24, 30–31.

11. *Balancing the National Interest: U.S. National Security Export Controls and Global Economic Competition*. Panel on the Impact of National Security Controls on International Technology Transfer. Committee on Science, Engineering, and Public Policy. National Academy of Sciences, National Academy of Engineering, Institute of Medicine (Washington, DC: National Academy Press, 1987).

eral reasons. First, it is hard to imagine the powerful defense industry of the Soviet Union and its allies dependent to any substantial degree on an influx of Western technology, including high technology. Second, the concept of dual-use technology may soon lose its meaning since most modern technologies become dual-use (see below). Third, the expansion of East-West restrictions may cause serious rifts between the U.S. and other industrialized states. Fourth, the expansion of restrictions causes a further deepening of the divisions between East and West, clearly contradicting natural tendencies and growing global needs.

The Generic Character of Technology

It is useful for our purposes to speak of civilian and military technology as separate entities. They have, in fact, different organizational and bureaucratic ramifications and different political connotations. Yet, as is often observed, this division is quite artificial and even arbitrary. There are, of course, examples of military technology having no or only rare applications to civilian life: nuclear weapons technology, stealth technology, MIRVs, high energy lasers. These examples do not indicate, however, that the basic technologies used in these weapons will never be applicable in the civilian sector, only that they are simply not usable now. Examples also may be found among those civilian technologies which, like computers or television devices, at first had no utilization in weapons systems but were later incorporated in full into military hardware. Thus, there is only one kind of technology encompassing all aspects of life and stemming from the same scientific and technical human effort. It is rational, purposeful human decision that divides technology into various categories and sectors.

This dividing line is never clear-cut, however; civilian and military technology are closely interlinked and overlapping. As one analysis has noted: "Technology does not sit neatly in little boxes marked 'military' or 'civilian.' An important element in the strains within the Alliance over sharing military-related technology is the widely held belief that advances in military technology translate directly into an advantage in civilian technology."[12] It is increasingly difficult to conceive of

---

12. Reppy and Gummot, "Economic and Technological Issues," pp. 17–18.

international cooperation in civilian technology while the military sector of this technology tends to divide states into opposing and mutually threatening groups. Military technology is subject to much more stringent rules than civilian technology, overpowering any economic interests which affect cooperation in the civilian domain. We may have it both ways—i.e., cooperation, on the one hand, and competition, on the other—but not for too long and only on a limited scale. The character of modern weaponry stresses the qualitative aspect. Increasingly, the basic components of weaponry—high-tech computers and materials, propulsion systems, communication gear—are also common ingredients of civilian products. It thus seems that technological developments are leading to a growing inability to separate the civilian from the military sector. If so, the chances for technological cooperation will be hindered. With no international action undertaken to decrease the function of military technology in security relations, civilian interests are bound to be subordinated. The reverse influence is much less probable; purely economic interests rarely prevail when vital security interests are invoked.

The generic character of technology should also be stressed in connection with the division of military technology into various categories of weapons. We are used to thinking of nuclear, conventional and space weapons as technologically distinct items. Nothing could be more misleading. These weapons are at present complex technological systems in which the only distinctive element is actually the final component—the weapon itself. The difference stems from the various physical processes of target destruction; all else is technologically the same for many modern weapons systems. If size and volume are eliminated from the analysis, then there is no difference between the guidance system for strategic or tactical missiles, between communications equipment used by a platoon commander or a national space command, between stealth materials used to cover strategic bombers and tactical cruise missiles.

Modern weapons systems, whatever their operational assignment and capabilities, have a common technological foundation and, in most cases, a common industrial base. Generally speaking, they cover the entire spectrum of possible directions; furthermore, all sectors of science and technology are mutually reinforcing. Achievements in one sector spread ever more quickly to the others. There are several material links between them, the most obvious being electronics, sensors and new construction materials. The increasingly generic character of modern weapons technology is clearly revealed by the fact that an increasing number of new but different weapons systems utilize exactly the same subsystems in their functioning. Any modern weapons system, space-based, strategic and tactical alike, operates according to a chain of action, each link of which has corresponding technical hardware—reconnaissance and location of target, communication of its location, elaboration of a decision and execution of the attack, launching a weapon or ammunition, delivery of the weapon, guiding and navigating the weapon on delivery, destroying the target, and, finally, assessing the damage. Each element of this chain can function in the service of a number of different weapons and on different levels of military operations. This is especially the case with reconnaissance systems, communication networks, data-fusion centers and means of transportation and delivery. Even if developed with only one specific weapons system in mind, the individual subsystem will eventually be utilized by some other weapons system or create a foundation for similar though slightly adapted subsystems. This interoperability of individual subsystems is the basis for the interoperability of the entire weapons system on a potential battlefield. The ultimate case of this phenomenon is found in multi-purpose weapons systems, be they dual-capable (nuclear and conventional), trilateral (nuclear, conventional, chemical), or interchangeable weapons systems (strategic and tactical weapons, e.g., some new cruise missiles). The generic character of modern weapons technology must be kept in mind when one analyzes the role played by technological factors in potential efforts at arms limitation and, more generally, in international security assessments.

III. Possible Solutions

If there is any chance to stem the tide of new conventional weapons technology, it has to be undertaken mainly by NATO and WTO states, with the participation of neutral and non-aligned European states. These states were the principal creators of the new weapons, and it is now incumbent upon them to restrict the danger they generated. The solution emerging from the above discussion seems inevitable. There is no better way to preserve stable and long-lasting European security, as well as technological development in both military and civilian domains, than arms limitation encompassing both nuclear and conventional weapons in both their quantitative and qualitative aspects. With nuclear weapons now under more intensive international scrutiny, there is an urgent need for a strong and purposeful effort in conventional arms control.

As has been pointed out, however, the matter of conventional arms limitation is so dynamic and so laden with various consequences, both political and military, that it is appropriate to ask whether the traditional approach will suffice. The traditional approach is composed of:

- negotiations concerned mainly with the numerical aspects of arsenals;

- an unrestrained technological modernization of arsenals under negotiation; and

- restricted cooperation in civilian technology (this element is not a part of the arms negotiations proper).

If one considers the four decades of experience using this approach, one may safely assume that before any agreement is reached, technology will overtake the process of accommodation. Some partial measures, if achieved at all in such circumstances, would not decisively alter the deterioration of European security. What could be achieved at most would be some numerical adjustments in manpower, enlarged confidence-building measures, perhaps some partial withdrawals of certain equipment from agreed-upon regions, and other such adjustment. All these measures would have a positive effect and should be pursued; however, they would not bring about stability that could be assured in time of crisis. They would not restrict further advancements of the technological arms race and they would not have any positive influence on civilian technological cooperation between states in the opposing alliances.

The ensuing dramatic expansion of conventional forces requires equally dramatic measures. They should be executed in addition to—and not as a substitute for—those undertaken in the traditional approach. These new objectives can be conceptualized, with some exaggeration and simplification, as 1) negotiating qualities instead of numbers; 2) tradeoffs between relevant weapons instead of similar kinds of weapons; 3) functions of military forces instead of potentials; 4) discussion of acquisition plans and processes instead of already existing weapons systems; 5) giving doctrines and structures an equal footing in negotiations with operational characteristics and deployments. Such an agenda is, obviously, extremely ambitious and seemingly above existing political possibilities. It would probably have to be carried out as a gradual process, aiming for a pace surpassing that of technology.

To make it plausible at all, the priority issue would have to be the most threatening and destabilizing weapons systems to be introduced in the next few years, such as medium- and short-range conventionally armed missiles, both ballistic and air-breathing, any upgraded air-defense missiles with the potential for BMD, and stealth aircraft and missiles. With this threat voided, other things are possible: gradually more and more open discussions of acquisition plans, indicating the true threat perceptions of the opponent; unilateral but reciprocal restraints on systems perceived as too threatening; restructuring of forces; and gradual redeployment of forces perceived as especially threatening. All these measures may be preventive arms limitations, separate from the structural and operational limitations embodied in the traditional approach. Preventive measures, such as limiting or cancelling planned weapons systems, would have a more direct impact on defense expenditures, military structures and doctrines than would limitations imposed on existing weapons.[13] One example of a preventive approach to arms control in Europe, particularly in central Europe, is the so-called Jaruzelski Plan, the four-point plan proposed by Poland on May 8, 1987 and further elaborated on July 17, 1987 and June 15, 1988. In this proposal, measures of gradual disengagement, beginning with the most destabilizing or threatening conventional weapons, are coupled with far-reaching transformation of the states' military postures. In addition, the plan proposed initiating a debate on the military doctrines of NATO and the WTO in order to secure the evolution of these doctrines toward less threatening, mutually acceptable military postures.[14]

The final point in the reasoning behind the call for a depar-

---

13. See Gale A. Mattox, "European Defense Policy and Arms Control," in Kelleher and Mattox, eds., *Evolving European Defense Policies*, p. 124; and Wolf Graf von Baudissin, "Arms Technology, Stability and the Arms-Control Deadlock: Part I," *New Technology and Western Security Policy*, *Adelphi Papers* No. 197 (London: International Institute for Strategic Studies, 1985), pp. 56–59.

14. For the text of the Jaruzelski proposal, see *Trybuna Ludu* (Poland), May 9–10, 1987, July 18, 1987, and June 16, 1988.

ture from the classical security policies in Europe is the realization that a further growth of the military technological competition, due to its increasingly strong connections to all aspects of technology, precludes the building of bonds between East and West in technological realms as well as in larger economic domains. While it may be argued that since so-called high technology engenders closer cooperation between allies there is no need for further enlargement of the market or resource supplies, not all allied states reap equal benefits from intra-alliance collaboration. Nonaligned and neutral states are by and large excluded. Restrictions on cooperation and collaboration with the socialist states and many developing states are likely to increase, disturbing the economic development of these countries.

It is not my intention here to favor the unconditional abolition of COCOM-type institutions; nor to propose some devious scheme to overcome their restrictions. In a world of fierce military competition, such institutions are understandable, if unwelcome and often apt to produce aberrant effects. The model offered here is a relationship in which a mutual exchange of technology would not directly endanger the other side by being utilized in new weapons systems. Where such utilization were foreseen, the most dangerous possible applications would be under restrictions and prohibitions derived from arms limitation agreements. In order to make this approach workable, it is necessary to negotiate future military technology and its application, and not to negotiate weapons systems already deployed. If such a safeguard system could be created for East-West technological cooperation, it would not jeopardize the basic security interests of either side. At the same time, military technology would not undercut future political, economic and technological cooperation.

It is clear from the above that no drastic reductions or limitations are foreseen in this approach, but such limitation would inevitably have to be undertaken either simultaneously or subsequently. The key is to buy time and gain trust in the future behavior of the other side and to streamline the options into those which are less destabilizing and threatening.

There is one more caveat to be observed, however, in regard to this approach to a new security arrangement in Europe. The vested interests of an army of engineers and technicians, industrialists and stockholders, and political leaders connected with local industries will not disappear because of some fine idea about arms limitation. It is indispensable, and highly profitable, to keep them busy and satisfied by providing them with alternative—that is, civilian—applications of the technology they develop. Only when the military is assured of the equal security of its country, only when the scientific-technical industrial vested interests are engaged in some other profitable activity, preferably close in technological level to that of the previous military production, will the whole approach have a chance to succeed. However, the proposal does not entail much conversion or industrial restructuring, at least not at the early stages of military technological limitations. It is sufficient to mention the various high-technology research and development programs undertaken in recent years by the West European states, the common outer space projects of various countries from East and West, joint aircraft productions, or the CMEA program on science and technology to perceive that limitations on high military technology need not be synonymous with technological retardation, conversion of production or bankruptcy. The ensuing freedom of international cooperation in various fields of science and technology would clearly benefit all participants. Moreover, there are a growing number of global challenges, both within and outside of Europe, which could invigorate entire branches of science and technology better than any military program.

Of course, one could pose a question: who is going to benefit more from the proposed drastic reorientation of military technology and the severe restrictions on some of its effects? Since some believe the West to be technologically ahead, the answer therefore would quite probably indicate the East as the sole beneficiary in economic terms, with both East and West gaining in their respective military security. However, the very same persons usually believe the East to be superior numerically and, moreover, to be quickly catching up qualitatively. This is, after all, an annual refrain during the budgetary debates throughout the Western parliaments. If so, then the answer to the question posed above does not hold water— restrictions would affect both sides. If not, and the East is only able to produce numbers without equal quality, then both sides have some currency to trade. As far as the long-term effects of the proposed approach are concerned, there is not the slightest doubt that the whole European community, together with both nuclear superpowers, would stand to gain. Several contentious issues of technology control, both among the Western states and between East and West, would vanish. It would be easier to accommodate the global regulations covering technological cooperation without detriment to national economic interests. Instead of counting on uncertain spinoffs from military research and development, industrialized states could reap the benefits of greater concentration on commercial goods. All countries could better satisfy growing social expectations of high living standards, expand their reserves of natural resources and better protect their natural environment.

5

The Impact of Military Technology on Conventional Arms Control in Europe

JOACHIM KRAUSE

Introduction

We have grown accustomed to hearing that technology has a major impact on our daily life, that it is a decisive factor in shaping modern economies, and that the future international competitiveness of a nation depends on its ability to cope with major technological challenges. Thus, it is no surprise that the impact of technology on arms control has emerged as a separate issue. With respect to strategic arms control this has always been appropriate, since technological innovations have caused rapid and profound changes in nuclear doctrine, strategy and force structures. Technological innovations have also changed the conditions under which stability has been defined and established. They have also impeded arms controls by fueling the nuclear strategic arms competition—through the introduction of MIRV technology, for example. But what is technology's effect on conventional arms control?

Many Soviet and East European authors, as well as a number of Western peace researchers and scholars, contend that the consequences of modern conventional weapons technologies are as far-reaching as those in the nuclear strategic field, and that decisive actions in the arms control area are imperative in order to address this issue directly. This paper will investigate the impact of conventional weapons technology on conventional arms control in Europe. It is based on the assumption that, in order to assess the impact of modern technology on *arms control*, one first has to consider its impact on conventional warfare and its likely consequences in terms of stability and arms race stimulation. For this reason, the paper consists of four parts. Part I is concerned with the impact new technologies have on conventional warfare; Part II discusses the impact of new technologies on conventional stability in Europe, especially central Europe; Part III addresses the question of whether new technology is a factor that drives the arms race; and Part IV deals with the consequences of technological change for arms control in the area of conventional forces and weaponry.

In the author's view, while technology has without a doubt an important impact on the conventional balance and on arms control, this effect is only measurable over the long term, and its immediate effect is often widely exaggerated. Furthermore, focusing future arms control efforts on technological issues would rather distract from the real stability and arms control problems pertinent in the central European conventional force balance. Moreover, talks or negotiations on technology issues would most likely be fruitless, since definitional and verification problems will probably be insurmountable.

I. Military Technology and Conventional Warfare

The impact of military technology on the way in which conventional war would be fought is often depicted in a rather simplistic fashion. Often, a few high-technology systems which are still in the making—or in the planning phase—are credited with having the capability of fundamentally changing the way wars are fought. It is true that in the past the advent of new technologies brought about fundamental changes in warfare. Tanks changed the nature of land warfare significantly; the emergence of fighter aircraft—especially bombers and fighter-bombers—had a profound impact on both land and sea warfare; and the introduction of radar had great significance during World War II. However, a closer look reveals that it was never simply the advent of one new item of technology that caused such changes. Tanks, fighter aircraft or radar systems were not single technologies but the result of an effort to bring together new and old and civilian and military technologies, in order to produce weapons or military systems of a new quality. It is the combination, therefore, of a number of technologies from different sectors and their reconciliation with existing technologies that results in changes in conventional warfare. This is true not only on the level of single new weapons systems; it is normally the synergistic effect of a number of technological innovations that profoundly changes the way conventional wars are fought.

New weapons systems alone, however, do not provoke changes in conventional warfare. Tanks and fighter aircraft had already shown up in World War I; neither of them was very decisive. It was not until their integration into new doctrines and force structures—doctrines and force structures developed in order to make optimal use of these new systems—that they could prove their real potential in warfare. Thus, if we speak about the impact of technology on conventional warfare, we must concern ourselves with more than merely new weapons systems or technologies. At least three things should be kept in mind:

- it is the synergistic effects of a variety of simultaneously emerging technologies, rather than single new developments, that may bring about significant changes in conventional warfare;

- the technical, managerial and political processes by which new technological developments are transformed into new weapons systems are as important as the invention of new technologies themselves;

- without the integration of new weapons systems into doctrines and force structures in a dialectic process, the impact of new technologies on conventional warfare will be negligible.

Military Technology and its Development: East-West Differences

In the process of developing and integrating military technology in their respective conventional force postures, East and West have manifested rather different features. The Western countries have been generally more innovative than the East with respect to the creation and development of new military-related technologies. This was mainly due to the broader and more sophisticated overall technological infrastructure of Western market-oriented industrial societies. One factor that contributes to this is that in Western countries the line between military and civilian research and development (R&D) is not as strictly drawn as it is in Eastern states. This provides the basis for a mutually beneficial cooperation. The Soviet Union has concentrated relatively much greater efforts on military R&D, as have, in part, its European allies as well. This made it possible for the Soviet Union to keep up with Western technology in a number of fields, and in some areas even to take the technological lead.[1] However, the lack of a technological-industrial base broad enough to keep up with all pertinent technologies, and the strict division between military and civilian technology have created the constant peril that the Soviet Union might fall behind in key technologies that are crucial for tomorrow's battlefields. In many areas (such as aircraft technologies, surveillance techniques, computer development and missile technology), they are behind the United States and often behind the other Western powers as well. On the other hand, Soviet military engineers have surprised Western observers time and again by developing workable low-technology solutions to problems that in the West were either seen as demanding high-technology answers or simply overlooked.

To complete this picture, one must take into account the asymmetries between East and West in the technical, managerial and political processes by which modern technologies are transformed into new weapons systems. Here, the West is at a slight disadvantage. There are various factors that render it difficult for Western countries to make full military use of their technological potential. First of all, militarily relevant innovations in technology occur not only in the United States but also in other Western countries, particularly France, Great Britain, West Germany, Italy and Japan. Although there is a grid of technology cooperation among these countries, there is no guarantee that technological innovations will be shared by all. Either security-related issues—for instance, the U.S. suspicion that technology secrets passed over to the Western European allies will soon find their way to the Soviets—or simple economic competitiveness have often prevented all NATO partners from profiting in time from new developments in military technology. A recent example is the withholding of stealth

---

1. John M. Collins, *U.S.-Soviet Military Balance 1980–1985* (McLean, VA: Pergamon-Brassey's, 1985), pp. 33–42.

technology from the West Europeans; such a technology, if made available for tactical aircraft in Western Europe, would decisively bolster the potential of NATO's air forces.

In addition, the Western countries waste their technological potential by duplicating R&D and arms production efforts that have already taken place—or are under way—in other countries. This is most visible in the fields of aircraft, missiles and electronics. The reason for this waste of money and resources is that most major Western industrial countries consider having their own capacity for arms development and production in these sectors essential on both economic and military grounds. Increased cooperation among various Western countries in the development and production of new arms has paid off only in part; armaments cooperation has often led to the production of new, extremely expensive high-technology systems of limited military use. Third, military technology is in most Western countries subject to political and budgetary constraints. Both the exotic way in which new technologies are sometimes presented and the often extremely high procurement costs encourage skepticism and cutbacks that diminish the military value of these new weapons systems.

On the Eastern side, these problems are virtually nonexistent, since only one country—the Soviet Union—possesses the technological and industrial infrastructure necessary to develop and produce a wide range of modern weaponry on a large scale. Political constraints such as those in the West are also nonexistent. It is prudent to assume that there might be budgetary restraints on the procurement of new high-tech weapons, but they are nowhere near as stringent as they are in Western democracies.

Finally, one should mention that there are also differences in the ability of East and West to reconcile new weapons with doctrines and force structures. Due to the lack of a unified Western military doctrine, most Western armed forces draw different conclusions—if often only slightly—about the impact of new technologies, and their military establishments usually put forward different demands for technologies to be developed. This adds to the already notorious lack of standardization and interoperability within NATO. Another relatively recent problem is the mounting public and scholarly criticism of new developments in doctrine and strategy designed to incorporate new technologies. The often fierce criticism uttered in the past few years against Air-Land Battle (a U.S. national doctrine) and

FOFA (Follow-on Forces Attack, a recently introduced component of NATO's strategy of flexible response) may serve to illustrate how strong such opposition can be and how much of an impact it may eventually have on the process of formulating doctrine and strategy. Again, on the Eastern side the picture is quite different. There is only one military doctrine in the Eastern camp—Soviet military doctrine. Due to its alleged scientific logic, the incorporation of military technology and the consequent improvements and refinements of doctrine, strategy, and force structures are paramount to Soviet military thinking. Whether or not Soviet doctrine, strategy and force structures are superior to their Western counterparts is not the issue in this paper. Nevertheless, it is important to point out that the Soviet military has much more leeway than its Western counterparts in defining doctrinal and strategic consequences of technological developments.

Returning to the initial question of the impact of technology on conventional warfare, it should be pointed out that the rapid and profound changes common in the area of strategic nuclear arms do not occur in the conventional arms field. Qualitative changes induced by technology usually take place in a more evolutionary process synergistically combining the effects of various technological developments. Thus, qualitative changes in conventional warfare do not take place overnight; they need some time to unfold. In this century, there have been only a few truly profound changes in conventional warfare introduced by new technologies: the mechanization of conventional warfare; the combined land-air warfare, as a consequence of the ability of fighter bombers and other aircraft to join ground battles; and the impact of nuclear weapons—especially tactical nuclear weapons—on conventional warfare. The next profound change, now underway, may result from what is called the "automation" of the battlefield, the combination of the effects of various trends towards automation and increased command and control possibilities in weapons technologies. There is, therefore, no need for arms control as a rapid cure against swift and sweeping changes in warfare as might be the case in nuclear strategic warfare. In looking for the possible consequences of future technology for conventional warfare and stability, we must focus instead on long-term evolutionary processes in a broad range of technological fields and their probable consequences, and we should consider in particular the impact of automation trends.

II. New Technologies and Conventional Stability in Europe

As a first step, the kinds of technology-induced changes that can be expected to emerge in the foreseeable future must be clarified. There is a huge body of literature on possible technological developments and their consequences for conventional warfare. It is beyond the scope of this paper to pay due attention to all or even some of the changes others have visualized—indeed, some of that literature rather resembles science fiction. Rather, it is sufficient to point out that both East and West agree that—due to the combined effects of a range of existing, emerging, or foreseeable technologies—conventional warfare in the 1990s and later may look different from that of today. These include developments in high-speed integrated circuits; radar evasion techniques; advanced software and algorithms; fail-safe/fault-tolerant electronics; artificial intelligence; high-speed computers; advanced materials and composites; high-density monolithic focal place arrays; fiber optics; high-energy lasers; anti-jam communications; precise navigation systems; real-time data fusion; battle-management systems; and other, as yet undeveloped technologies.[2] All of these may make possible the development of weapons systems and force postures that allow fast, coordinated and concentrated attacks on a growing number of both fixed and time-critical targets—at greater depth, with steadily improving accuracy, and with increasingly effective conventional munitions.

In the words of a leading U.S. defense official, expectations run high with respect to these new technologies:

> By judiciously using the emerging technologies, we can improve the capability of NATO forces to acquire distant targets; obtain surveillance and intelligence information; improve the communication between remote sensors, command centers, and weapons; utilize advanced methods for data fusion to consolidate, analyze, and disseminate battlefield status and intelligence information; increase the effectiveness of combined air and land anti-air, anti-armor and counterbattery tactics; and enhance the capability to interdict enemy rear areas through remote target

---

2. Richard D. DeLauer, "Emerging Technologies and Their Impact on the Conventional Deterrent," in Andrew J. Pierre, ed., *The Conventional Defense of Europe* (New York: Council on Foreign Relations, 1986), p. 51.

acquisition, standoff weapons attack, and long-range attack on enemy reserve forces, resupply areas, airfields and depots.[3]

The words of a leading Soviet defense official reveal a vision of similarly revolutionary change in conventional weapons technology:

> Rapid changes in the development of conventional means of destruction and the emergence in the developed countries of automated reconnaissance-and-strike complexes, long-range high-accuracy terminally guided combat systems, unmanned flying machines, and qualitatively new electronic control systems make many types of weapons global and make it possible to sharply increase (by at least an order of magnitude) the destructive potential of conventional weapons, bringing them closer, so to speak, to weapons of mass destruction in terms of effectiveness. The sharply increased range of conventional weapons makes it possible to immediately extend active combat operations not just to the border regions, but to the whole country's territory, which was not possible in past wars. This qualitative leap in the development of conventional means of destruction will inevitably entail a change in the nature of the preparation and conduct of operations, which will in turn predetermine the possibility of conducting military operations using conventional systems in qualitatively new, incomparably more destructive forms than before ... There is a sharp expansion in the zone of possible combat operations, and the role and significance of the initial period of the war and its initial operations become incomparably greater.[4]

Both alliances are currently endeavoring to increase their knowledge and skills in these technologies, to develop new military systems, and to incorporate them into their respective doctrines and force structures. The pertinent question is whether this will lead to more or less stability. Contrary to the claims of those who generally attribute a highly destabilizing effect to new technologies, this question is not easy to answer. These newly emerging, existing or foreseeable technologies will not of themselves make for more or less stability. In gaug-

---

3. Ibid., p. 50.

4. Marshal N. V. Ogarkov, Interview with *Krasnaya Zvezda* from May 9, 1984, p. 2-3; English quotation from *FBIS Daily Report Soviet Union*, Foreign Broadcast Information Service, May 9, 1984, p. R.19.

ing their impact on stability, their specific impact on the further development of the strategies, doctrines, and force structures of the North Atlantic Treaty Organization (NATO) and the Warsaw Treaty Organization (WTO) must be assessed. In order to arrive at that point, we first have to determine what we understand about the stability of conventional forces in Europe and the relevance of strategies, doctrines and force structures for this stability.

Stability and the Role of Doctrine

The notion of stability is an offshoot of strategic nuclear arms control. It is difficult, however, to apply the concept of stability to the area of conventional forces. Nuclear weapons, due to their unique physical characteristics, lend themselves as weapons of deterrence; strategic nuclear stability can be built around the concept of mutual deterrence, as was the case with the concept of mutual assured destruction (MAD). To define stability on the level of conventional forces, however, is hardly feasible. Conventional weapons do not have the ultimate character of nuclear weapons; taken alone, conventional forces will influence the risk-taking on the side of a potential aggressor, but they cannot serve by themselves as the backbone of a deterrence strategy. History has shown that even a conventional equilibrium may not be sufficient to deter a potential aggressor from initiating a war.

In other words, it would be pointless to try to define stability at the conventional level. There is, however, room for a stability-oriented concept for conventional forces, as long as they are not viewed in isolation but as part of the overall deterrence balance in Europe—a combination of nuclear and conventional forces where the main peace-preserving effect results from nuclear forces. Here one may speak of stability as the absence of conventional attack options that undercut mutual deterrence, such as surprise attacks or attacks with short warning time.

NATO and the WTO have responded differently to the idea of stability as defined above. Since the development of the nuclear stalemate in the 1960s, the military efforts of both sides have been characterized by a greater emphasis on conventional warfare. This has led to major changes in doctrines, strategies and force structures. Nonetheless, there is a clear difference between the shift in force postures of NATO and the WTO:

whereas NATO has confined its preparations to the improvement of its ability to fight a defensive war, the Soviet Union and its allies have done the opposite—they have improved their already offensively-oriented force posture. NATO's efforts were concentrated on enhancing flexibility in a future war that would be fought defensively on its own territory from the beginning—and in which early war termination would be the paramount objective. The respective efforts of the Soviet Union and its allies were aimed at providing the military capabilities, using conventional means only, to launch a preventive strike against nuclear delivery means and related facilities for NATO command and control and to mount a *blitzkrieg*-type offensive in Western Europe. As Michael MccGwire recently put it, "such operations would make NATO's resort to nuclear weapons much more difficult, and, even if the Soviets were not fully successful, NATO's nuclear capability would be greatly reduced and the escalatory momentum would be lessened."[5] One may argue whether these efforts were based on a master plan to sap the very foundations of NATO's deterrence/defense posture in Europe, or whether it was a more or less normal development, since Soviet military thinking generally prefers offense over defense. The fact is, however, that by acquiring the capability to successfully launch a non-nuclear *blitzkrieg*-type attack on Europe, the Soviet Union has been striving for a conventional option that undermines the foundation of stability in Europe.

When we look in this way at both NATO's and the WTO's force postures in Europe, we discover that the impact of military technology on stability is dependent upon which doctrines, strategies and force structures the technology is intended to improve. This is true not only for new and emerging technologies. The introduction of such seemingly defensive systems as mobile anti-aircraft missile launchers, for instance, can significantly add either to the defensive or to the offensive potential of an army. While they would help a defender in becoming more flexible against a variety of air threats, they would also help an offense-oriented force to enhance its mobility, thus adding to the speed of advance. In examining the WTO military posture, it is evident that the actual impact of these new technologies on stability could be very severe, since

---

5. Michael MccGwire, *Military Objectives in Soviet Foreign Policy* (Washington, DC: The Brookings Institution, 1987), p. 338.

they might help the Soviet and East European forces make significant progress in improving their non-nuclear offensive capability.

If, as was outlined earlier, the Soviet concept of a non-nuclear war in Europe were linked to the possibility of rendering impotent NATO forward defense, deliberate preparations would be directed at achieving a maximum of military success in the earliest phase of war, without the initial use of nuclear weapons. This could only be accomplished through the optimal use of air power and weapons reaching far into the depths of Western defense, in order to disrupt the buildup of Western forward defense forces and destroy as many nuclear delivery means as possible in the very first hours and days of war. Only the success of such an initial air operation—which must also be an anti-air operation—could provide the conditions necessary for breaking through NATO's forward defense with ground forces within the first days of the attack. Such a concept thus calls for a capability to hit as many fixed and time-critical targets in the whole depth of Western defense as possible within the early stages of a war.

With its current air force and missile inventory, the Soviet Union would have serious difficulties conducting an initial strike sweeping enough to substantially downgrade NATO's conventional buildup and destroy NATO's nuclear delivery means. However, the new emerging and foreseeable technologies of conventional warfare might change that picture decisively. Future developments in surveillance, target acquisition, and C^3I technologies; improvements in the range of artillery and the "smartness" of shells; air-launched stand-off weapons; radar-evading techniques; and increased ballistic missile accuracy will most likely give the Soviet Union in the 1990s a much better capability to launch such an initial air operation against NATO's military installations in the Western part of central Europe. Such an air—and anti-air—operation might never succeed in defeating NATO's forces alone, but it could pave the way for a consequent offensive by huge ground forces. Thus for the Soviet Union to succeed in acquiring, developing and introducing those new technologies would be the most destabilizing scenario for European security since the 1950s.

The same detrimental consequences, in terms of stability, would not result if NATO were to acquire weapons systems based on these new technologies, because of the defensive character of NATO's strategy which outrules large-scale offensive operations by ground forces. Current developments in the context of FOFA do not militate against this defensive character. They are aimed at denying to the WTO the prospect of a successful non-nuclear attack on Western Europe, thus contributing to the maintenance of stability. Some observers, however, have contended that many of those weapons systems envisaged for FOFA—such as, for instance, long-distance weapons—could lead to a decrease in stability because they give NATO a preemptive capability. Even if NATO does not intend to make use of that preemptive capability, it could at least create such an impression on the Soviet side. Such scenarios are, however, highly artificial and militarily absurd. As long as NATO does not possess those forces, force structures, doctrines and strategies that would enable the Western alliance to exploit the weakening effect of such a preemptive strike by launching a broad and fast invasion by its ground forces, there is no rationale for the West to strike preemptively with these envisioned FOFA weapons. Again we return to the point that strategies, doctrines and force structures are the real factors determining stability or instability.

To sum up, although new technologies alone—existing, emerging or foreseeable—do not pose a problem in terms of conventional stability in Europe, they could aggravate the already existing stability problems posed by developments in Soviet strategy, doctrine and force structures. Thus there is no need to address weapons technology as an issue of its own in arms control. Instead, arms control negotiations should directly address the stability problems resulting from the current Soviet military concept of non-nuclear offense and its further evolution in the light of impending technologies. The recent interest of the Soviet Union and its allies in discussing military doctrines indicates that this point of view has also won support among Eastern policymakers and analysts, although we are still far from the point where serious negotiations on doctrines or strategies can begin.[6]

III. New Technologies and the Arms Race

One argument repeatedly heard in favor of making weapons technology the centerpiece of arms control is that military tech-

---

6. Joachim Krause, *Prospects for Conventional Arms Control in Europe*, Occasional Paper #8 (New York: Institute for East-West Security Studies, 1988).

nology drives the arms race, thus frustrating arms control efforts. Again, this notion stems from the nuclear strategic arms competition; in this area it may be reasonable to view technology in such a way. With respect to conventional forces in Europe, however, this argument is flawed because it is hard to find an arms race on the level of conventional forces in Europe. The concept of an "arms race" points to the fact that offensively oriented military preparations in peacetime by competing powers might set off a spiral of force buildups on both sides, even when these military preparations arise from defensive political intentions. The decades before World War I are often referred to as a typical example of an arms race. The nuclear competition between the two superpowers manifests strong elements of an arms race as well.

The conventional armaments competition in Europe, however, has never borne the characteristics of an arms race. Rather, it might best be described as a competition in which one side constantly strives for superiority and improvement of its offensive capacities, in order to secure swift and sweeping victory in a war that is to be dominated offensively from the outset, while on the other side, a military coalition tries to retain forces which will be able to negate the first side's prospects for such a victory, without necessarily trying to match its military efforts quantitatively and qualitatively. That there is no arms race on the conventional level in Europe may be demonstrated simply by comparing force inventories on both sides and their respective changes. There are already conspicuous numerical imbalances between East and West. But if one looks at the increases in weaponry over a given time on both sides, the difference is startling. Between 1965 and 1980, for instance, the Warsaw Pact introduced on an average four times as many major weapons as NATO in the central European region, and in some categories even six or eight times as many.[7] If there were an arms race, the increases in weaponry would be at least similar in quantities or qualities, but it is obvious that the driving force behind the increase was the one side that followed an offense-oriented strategy—the Warsaw Pact.

Some observers contend that the relatively higher standard of technology in the West is a factor that drives the arms com-

---

7. Philip A. Karber, "To Lose an Arms Race: the Competition in Conventional Forces Deployed in Central Europe 1965–1980," in Uwe Nerlich, ed., *The Soviet Asset—Military Power in the Competition Over Europe* (Cambridge: Ballinger, 1983), esp. p. 81.

petition in conventional forces. This argument is true in the respect that it was the West rather than the East which introduced new technologies. It is, however, incomplete, since the East has, as a rule, introduced more quantities of weapons of higher technological standards than the West, and it has always tried to make up for actual or alleged Western technological leads by increases in numbers. Thus, it would be wrong to place responsibility for the armaments competition on the Western striving for technological solutions to problems that were raised by the WTO bid for conventional superiority. As has been demonstrated above, the main driving impulse for the conventional armaments competition in Europe has come from the Eastern side, since the offensive orientation of their conventional force posture demands superiority and concealment—and both are elements which contradict the logic of arms control thinking and stability-oriented policy.

IV. Technology as an Arms Control Issue

In returning to the initial question of the impact of technology on arms control and the likely consequences for arms control policy, there are two remarks that are pertinent. First, it would be wrong to let military technology become the focal point of arms control. The most important stability problems in the conventional field in Europe derive from the offensive character of Soviet and East European military preparations. Their strategy, doctrine and force structure, and the concomitant drive for superiority in manpower and equipment which is necessary to meet this highly ambitious military concept, constitute the main stability problem, not the purportedly destabilizing character of new technologies. Second, these qualifications notwithstanding, technology is one issue for conventional arms control, but it has to be addressed in a way that corresponds to its importance relative to other factors. It is in this framework that certain oft-mentioned arms control options related to technology issues have to be discussed.

The first option would be to single out special military technologies which are extremely destabilizing and to negotiate a ban on them. At first glance this idea is intriguing, but it would be very difficult to find weapons systems that lend themselves to such an endeavor. As was mentioned earlier, no weapons system is either purely defensive or purely offensive—its orientation depends on which forces will use it in which environment and for which purposes.

Another option might be to bring about an agreement between East and West to forego the development and deployment of military technologies that will lead to "revolutionary" changes in conventional warfare in the foreseeable future. This sounds reasonable at first glance, since these new technologies will—when they are integrated into an offensive strategy— cause a substantial increase in instability. Again, going into the details of such a deal reveals some insurmountable obstacles. First, there is an asymmetry hidden in such a proposal that mainly favors the East. NATO countries have never tried to match the immense Soviet and East European conventional buildup with equal numbers but with superior military technology, which at least provides a good chance to deny the East any successful swift-attack option. If the renunciation of new weapons technologies should become the subject of an arms control agreement, it could lead to destabilization, since the numerical imbalances will be untouched and the prospect for successful conventional attack options will rise for the WTO. The second argument against such an approach relates to definitional and verification problems. It will be extremely difficult to find parameters that lend themselves to such an approach. It is not that easy to define some technologies or weapons systems and then outlaw them. The most natural way might be to focus on delivery systems; all other parameters would be too difficult to define and verify. The banning of delivery systems, however, would make sense only if one outlaws whole categories of weapons, among them aircraft, missiles and artillery munition dispensers. The ban on all ballistic missiles proposed at the 1986 Rejkyavik summit would have had such an impact, since it would have excluded the development of highly precise missiles which could be tipped with conventional charges. But this proposal turned out to be too utopian. Ultimately, the banning of certain kinds of delivery systems would lead to the elimination of the most common delivery systems that contemporary armed forces possess. Thus, it is hardly realistic to expect that it is possible to ban all these new technologies from coming into existence.

Another arms control option could be to look for tradeoffs between Western technology leads and Eastern numerical advantages. The logic behind this is that if the West balances the numerical superiority of the Warsaw Treaty with better technology, the West could offer to forego the development or deployment of certain new technologies when the WTO is prepared to offer adequate reductions in numbers. It is impossible to trade continually in the renunciation of the development of new technologies, however, since a technological lead usually lasts only a few years, at most a decade, and without continued R&D efforts the overall technological lead of the West would dwindle within a few years, leaving the WTO free to redeploy its previously withdrawn forces. The only possible way to establish a link between Western technological leads and Soviet numerical advantages would be to withdraw or withhold certain high-technology weapons from the European theater as long as the Soviet Union agrees to withdraw equally important weapons in a significant number from the central European region. One example could be to withdraw a certain amount of air-launched anti-tank weapons—such as the AGM-65 Maverick—in exchange for the withdrawal of a certain number of Soviet tanks and other armored vehicles from the Group of Soviet Forces in Germany, or to link agreements on ceilings for those weapons systems in a given zone. Such an endeavor would not be of great risk for the West, since in case of a violation the redeployment of those anti-tank missiles would be easy. Due to the differences in weight between tanks and anti-tank missiles, the differences in geography could be outweighed in that case. Whether or not such a swap will be negotiable and under what terms remains an open question. The main problem about such a deal is that it does not address directly the underlying differences between NATO and WTO in doctrine, strategy and force structure.

A fourth arms control option would be to reduce the element of insecurity over the military R&D efforts of the other side with confidence-building measures (CBMs). These CBMs must give assurance that no new developments are under way that are able to transform the way conventional wars are fought. This idea, again, sounds intriguing, but its implementation will most likely be disappointing. As was mentioned earlier, "revolutionary" changes in conventional warfare are as a rule the consequence of the combined effects of evolutionary changes that take decades to materialize fully and which have to be implemented into new doctrines, force structures and strategies in order to develop their whole potential. Thus, the effect of such CBMs will be minimal. Conversely, there could be a furthering of confidence if both sides would issue enough information about current R&D efforts in the military fields. Here, a lot of steps have already been taken in the West, where information on weapons developments is readily available in professional journals, legislative deliberations and public hearings. If the Soviet Union and its allies are seriously interested in a dialogue on technology, it would be desirable to have the same kind of openness on the Eastern side.

Thus, technology does not seem to be a serious issue for arms control in the field of conventional forces in Europe. There is some leeway for a broadened consideration of technology problems, and there should be a deeper understanding of the fact that pending technological developments will aggravate the existing instabilities on the conventional level. But it would be wrong to make military technology the centerpiece of arms control efforts.[8] This would distract from the real stability problems we are facing in the conventional area and make arms control an even more arduous endeavor.

---

8. See Andrzej Karkoszka's chapter in this volume.

6

The Impact of Military Technology on the Arms Race: Armaments Dynamics in the Nuclear Age

MAREK THEE

The Unremitting Arms Race

In the four decades since World War II, we have been reassured daily by promises of disarmament and arms control. Yet the stocks of nuclear and conventional weapons continue growing without interruption. Nobody knows how to use nuclear weapons in an operational way without inflicting suicidal harm to humanity, the perpetrator included. As late as 1979, even Henry Kissinger remarked on the tactical nuclear buildup in Europe: "We never had a comprehensive theory for using theater nuclear forces . . . . We had no precise idea what to do with them."[1] And today, the nuclear arsenals of our globe are filled with 45,000 to 56,000 nuclear warheads, strategic and tactical,[2] even though a small number of heavy warheads would suffice to destroy the world as we know it.

How did we reach this point? What is the driving force behind this folly?

---

1. Henry A. Kissinger, "The Future of NATO," *Washington Quarterly* 2, No. 4 (Autumn 1979), pp. 5, 7.

2. See R. W. Fieldhouse, "World Nuclear Weapon Stockpiles," in Marek Thee, ed., *Arms and Disarmament: SIPRI Findings* (Oxford & New York: Oxford University Press, 1986), p. 77.

There is no simple answer to these questions. However, closer study of the history of modern weapons systems indicates that in this heedless arms race, a critical role has been played by the unrestrained forward course of modern military technology.[3] Indeed, this has been explicitly confirmed by two top professional authorities, former U.S. Secretary of Defense Robert McNamara and Nobel Prize winner Hans A. Bethe:

> The 25,000 [nuclear] warheads that each nation [the U.S. and USSR] possesses did not come about through any plan but simply descended on the world as a consequence of continuing technological innovation.[4]

The arms race, of course, is a phenomenon with many causes. Modern military technology, perceived as an extraordinary force multiplier, provides the "push power" for armaments.

Of the various factors that fuel today's arms race, two main determinants stand out: the competition in military technology and military-political-economic vested interests. The latter, termed by former U.S. President Eisenhower the military-industrial complex, includes the military and military industry, who are interested in fostering their economic and sociopolitical privileges in society, as well as the state political bureaucracy, which is interested in using military power as an instrument of policy and diplomacy. In addition, there are also the large and influential scientific-technological establishments of major powers engaged in military research and development (R&D) with interest in promoting their particular professional concerns.

A generic circular relationship exists between the race in military technology and the various group interests mentioned above. Combined, these make up the political economy of armaments—the way the arms flow is handled on military, industrial and administrative levels, from the technological inception and maturation of weapons systems to their production, acquisition and deployment. These two determinants are

---

3. Marek Thee, *Military Technology, Military Strategy and the Arms Race* (London & Sydney: Croom Helm, and New York: St. Martin's Press, 1986), pp. 14–41.

4. Robert McNamara and Hans A. Bethe, "Reducing the Risk of Nuclear War," *Bulletin of Peace Proposals* 17, No. 2 (1986), p. 127.

mutually reactive. In the light of the specific impact of the postwar revolution in military technology, however, the "push power" of modern military technology emerges as a prime agent of the contemporary arms race.

Ever since World War II, the competition for technological superiority and for increased war-fighting and war-winning operationalization of nuclear and conventional weapons has dominated the East-West military environment. This is epitomized by the unremitting exertion of military R&D. As stated by Edward Teller, "it's not the deployment of weapons that counts, it's what goes on in the laboratories."[5] Such national laboratories in the United States as Livermore and Los Alamos compete fiercely to develop new generations of weapons "beyond stated requirements . . . in part because their funding and recruiting success depends on finding new challenges."[6]

In fact, military laboratories act to preempt the political decision-making process. Here it is worth recalling the lessons drawn by Herbert York, former director of the Lawrence Livermore Laboratory, in the context of the decision to proceed from the development of the atomic bomb to production of the (super) hydrogen bomb. Despite initial soul-searching debates in the U.S. scientific community about the wisdom and morality of further development of nuclear weapons, the push effect of the new military technology proved irresistible. York writes:

> This particular episode, like the history of the super [bomb] itself, can be seen as an illustration of how what Secretary of Defense McNamara called technological momentum can determine the course of the arms race. The possibilities that welled up out of the technological program and the ideas and proposals put forth by the technologists created a set of options that was so narrow in the scope of its alternatives and so strong in its thrust that the political decision makers had no real independent choice in the matter.[7]

---

5. Norman Moss, "Sunday with Edward Teller," *The Listener* 13 (June 1985), p. 14.

6. Fred Hiatt and Rick Atkinson, "Lab Creating a New Generation of Nuclear Arms," *The Washington Post*, June 9, 1986.

7. Herbert York, *The Advisors: Oppenheimer, Teller and the Superbomb* (San Francisco: Freemen and Co., 1976), p. 11.

The thrust of modern military technology has set the stage and critically positioned the arms race, vertically and horizontally, in quantity and quality, in nuclear and conventional weapons. This revolutionary technological momentum has had a determining impact on the course of the arms race. The incorporation of nuclear weapons into U.S. and Soviet military strategy, stimulated by the perception of these being the "ultimate" weapons, has helped to harden the positions of the two powers into long-term confrontational postures. It is difficult to overestimate the role of modern military technology in shaping strategy and its decisive influence on international politics.

Seen from this angle, the expansion of the arms race into outer space, propelled by the U.S. Strategic Defense Initiative (SDI), marks a significant new turning point—a qualitative leap into a new phase of the arms race. In terms of breakthrough weapons development, the underlying aim is to complement existing arsenals—both nuclear and conventional—with futuristic and exotic weapons, thus reaching out for new frontiers of war-fighting capabilities. Unprecedented massive investments in new military technology are likely to yield new and aggressive accomplishments in esoteric weaponry. Such a situation will obviously be fraught with military destabilization and increased international tension and insecurity.

Armaments Dynamics

Twenty years ago, at the peak of the unfolding of new weapons systems—the diversification of nuclear arsenals, the deployment of intercontinental land- and sea-based ballistic missiles, the launching of military satellites and the development of the independently targetable reentry vehicles (MIRVs)—then Secretary of Defense Robert McNamara admitted:

> There is a kind of mad momentum intrinsic in the development of nuclear weaponry. If a system works—and works well—there is strong pressure from all directions to procure and deploy the weapon out of all proportions to the prudent level required.[8]

---

8. Remarks before United Press International Editors and Publishers, San Francisco, September 18, 1967, *Department of State Bulletin*, October 9, 1967.

In a nutshell, this statement encapsulates a cardinal feature of contemporary armaments dynamics: the previously mentioned "push power" of the rapidly evolving modern military technology. This then interacts in a circular way with the political economy of armaments as manipulated in the competitive process of production, acquisition and development of new weapons systems by the economic-military and bureaucratic-political "pull power" of the military-industrial complex. Consequently, the technological momentum emanating from the military laboratories imposes itself on the political decision-making process.

As a corollary, the arms race between the major powers is intensified. The action-reaction and overreaction mechanism is activated. One side's advances in military technology, even when of a defensive nature, are perceived and interpreted by the other side as having political and military intent. Acting on worst-case assumptions, the other side produces a reaction "out of proportion to the prudent level required." The inner armaments momentum is then reinforced by external arms stimulants. A vicious circle of action-reaction and overreaction is set into motion. Underlying the *pull* of the political economy of armaments, as a key agent of the arms race, is the technological *push*.

Telling examples of this process are offered by the emergence and deployment of two crucial weapons systems of our times: MIRVs and strategic cruise missiles. According to Herbert York, who was also chief scientist of the U.S. Advanced Research Project Agency (ARPA) at the Office of the U.S. Secretary of Defense:

> The MIRV program was almost entirely technologically determined in the sense that the key decisions were made by technologists who were either attempting to solve problems posed by nature, or responding to their perceptions of the technological challenges posed by Soviet missile and space programs.[9]

And the Brookings Institution study on the origin of cruise missiles states:

> Cruise missiles have evolved without a well-defined conception of why they were needed, and without an assessment of their

---

9. Herbert York, *The Origins of MIRV*, Stockholm International Peace Research Institute, Report No. 9 (August 1973), p. 22.

full implications. The programs illustrate how U.S. research and development sometimes operate independent of the policy process.[10]

In the case of the Strategic Defense Initiative as well, the real inspiration came from some hawkish sections of the scientific-technological community, most prominently represented by Edward Teller, the renowned "father" of the hydrogen bomb. Teller had four meetings with President Reagan before the so-called "Star Wars" speech of March 1983.[11] A week later, on March 30, Teller wrote in *The New York Times*:

> Today, a wide range of good and ingenious technical plans, ranging from simple to extraordinarily complex, challenge the widespread opinion that practical defense cannot be obtained. Mr. Reagan did not lightly accept the idea that these can be made to work. He wanted to know a vast number of details. He asked questions of his science adviser, George Keyworth, and of many other scientists, myself included. He decided that something must be done.

The push power of military technology ignited the pull power of the political economy of armaments. The initiation of SDI represents the latest culmination to date of the internal propulsion in and infatuation with military technology.

The Mode of Operation of Military R&D

For a better understanding of the role played by modern military technology in stimulating the arms race, we need insight into the workings of military R&D—its structure and operational imperatives. After World War II, military R&D expanded rapidly in the wake of the emergence of nuclear weapons and the explosion of modern military technology. Whereas prior to World War II military R&D consumed on the average less than 1 percent of the military expenditure of major powers,[12] it has

---

10. Richard K. Betts, ed., *Cruise Missiles: Technology, Strategy, Politics* (Washington, DC: The Brookings Institution, 1981), p. 1.

11. William J. Broad, "Reagan's 'Star Wars' Bid: Many Ideas Converging," *The New York Times*, March 4, 1985.

12. *World Armaments and Disarmament: SIPRI Yearbook 1974* (Stockholm: Almquist & Wicksell, 1974), p. 127.

in recent years expanded to comprise 11 to 13 percent of these expenditures.[13] Military R&D is the fastest-growing item of military spending, now receiving a far greater share of government expenditure than civilian research and development—over 70 percent in the United States.[14] Similar estimates—some even higher, given the efforts to catch up and surpass the U.S. performance—have been cited for the Soviet Union.[15] Such expansion of military R&D into almost all fields of basic and applied research means that today the military holds a controlling position in almost all contemporary R&D endeavors. A new order of magnitude in harnessing modern science and technology for the arms race has come into existence. In the United States, about 40 percent of scientists and engineers are on the payroll of military R&D. On a global scale, military R&D employs today approximately 750,000 to one million of the best-qualified engineers and scientists, with a budget of about U.S. $100 billion annually.[16] This in itself must necessarily have a dramatic effect on the arms race.

Beyond the quantitative dimensions—its vast resources and penetration into all domains of scientific-technological research—military R&D exerts a potent impact on armaments through its structure, institutional setup, and operational imperatives. Though dispersed in thousands of laboratories, university research centers and industrial plants, military R&D is a well-knit enterprise, mission-oriented, committed and functional. Among the features of its inner dynamics and perseverance we should note: a) the long lead times required for the conceptualization, prototype production, repeated testing and production of new weapons systems. These extend on the average to 10 to 15 years, generating a long-haul technological push beyond the lifetime of single government administrations. Thus, military research projects tend to pass from one administration to another, unaffected by political fluctuations (a good

---

13. Mary Acland Hood, "Military Research and Development," *SIPRI Yearbook 1985* (London & Philadelphia: Taylor & Francis, 1985), p. 289.

14. Franklin A. Long, "Government Dollars for University Research," *Bulletin of the Atomic Scientists* 42, No. 3 (March 1986), p. 289.

15. See Marek Thee, *Military Technology, Military Strategy and the Arms Race*, p. 105.

16. Ibid., pp. 107–108.

example here being the history of the U.S. MX missiles); b) the follow-on imperative, which—as a matter of professional routine, expedience and scientific curiosity—requires that each achievement in military technology be followed up and upgraded with further efforts at product improvement and "modernization," and that all new weapons of offense be complemented with weapons of defense (and *vice versa*) on the assumption that the adversary has or will acquire similar or even more advanced arms; c) the "block-building" confluence of technologies, by which different, initially unrelated technologies meet in a cross-fertilization process; and d) worst-case analysis and planning, by which, in the atmosphere of deep secrecy surrounding the arms race, "conservative prudence" dictates added zeal in overdesign and overplanning so as to preempt the adversary. It actually instills excess in the minds of scientists and technologists as well as military-political planners.

All these phenomena and operational imperatives of military R&D converge and intertwine to invigorate, routinize and solidify the technological momentum, infusing permanency and continuity in the armaments effort and projecting it far into the future. Once undertaken, R&D projects on new weapons systems acquire a life of their own, with inertial firm commitments for a long time to come. They will continue undisturbed by the outer political environment, be it the state of arms control negotiations or change of administrations. In this way military R&D has a pervasive impact on the course of the arms race: its push power has imposed itself on the political process.

Such military-technological momentum necessarily represents a cause of permanent concern in the United Nations. The UN Comprehensive Study on Nuclear Weapons has remarked:

> The development of nuclear-weapon technology has created an important dimension in the arms race. It is clear that in many cases technology dictates policy instead of serving it, and that new weapon systems frequently emerge not because of any military or security requirement but because of the sheer momentum of the technological process ....[17]

---

17. UN Study Series 1, *Comprehensive Study on Nuclear Weapons* (New York: United Nations, 1981), para. 493.

To this the UN Study on the Relationship Between Disarmament and Development adds:

> One of the most conspicious distinguishing features of the military scene since the Second World War has been the extraordinary rapid rate of change in weapon technology. It is this feature of the post-war arms race that is primarily responsible for the unique intensity of this race.[18]

These mechanisms and interactions among the workings of military technology, the political economy of armaments, and the bilateral stimulation between contending major powers form essential elements of contemporary armaments dynamics. However, they do not exhaust the complexities of this momentum.

Of critical consequence is the impact of the doctrine of nuclear deterrence. Its underlying "balance of terror" theory infuses threat and intimidation into the international system, thus additionally fueling the arms race. It actuates mutual suspicion and mistrust. Essentially, "balance of terror" is a convoluted way of describing and stimulating the arms race. In an overarmed world, obviously no power takes the concept of military balance for granted. There is an immanent and irresistible inner pressure, stimulated by the forward course of military technology, to always "do better" than the adversary. This propensity eventually channels into an intensive drive for military superiority, locking East and West into an open-ended arms race.

Thus, the mode of operation of military technology, the aggressive involvement of vested interests, the confrontational interaction between the major powers and the psychological-behavioral and doctrinal superstructure all meet to buttress each other. The end result is the powerful "mad momentum" of the arms race.

Pairing Offense with Defense

The focal point of the arms race since the emergence of the nuclear age has centered on efforts to make nuclear weapons

---

18. UN Study Series 5, *The Relationship Between Disarmament and Development* (New York: United Nations, 1982), para. 146.

usable and operational for nuclear war-fighting purposes. Nuclear weapons have been miniaturized and put on long-distance and intercontinental delivery vehicles; they have been perfected in accuracy, speed, maneuverability and target acquisition. Their constant refinement has produced increasingly sophisticated strategies of "escalation control and dominance," "counterforce," "countervailance" and "prevailance" in a protracted nuclear war.

Nonetheless, the cataclysmic destructive potential of nuclear weapons has marred and outweighed their useability. The use of nuclear weapons spells incalculable ruin, transcending any of the values and aims for which wars have traditionally been fought. Hence the pursuit of third-generation nuclear and exotic weapons to complement the nuclear potential both in operational utility and strategic employment. It reflects the aspiration of nuclear powers to extricate themselves from the straitjacket of mutual nuclear vulnerability.

Thus, underlying the race for exotic weapons is a double motivation. The first is to enhance the war-fighting capabilities of existing nuclear and conventional weapons. The second is to arrive at a nuclear-exotic war-winning capability by pairing, in a classical Clausewitzian way, the offensive potential of the conventional and nuclear arsenals with the expected defensive capability of exotic weapons. Actually, in his Star Wars speech of March 23, 1983, President Reagan admitted the following about defensive systems: "if paired with offensive systems, [they] can be viewed as fostering an aggressive policy."[19] Whatever the assurances that this was not intended, the aggressive nature of mixing offense with defense remains an objective reality and is unequivocally apprised as such in any strategic theory. This combination of offense and defense, as underlined by Leslie H. Gelb in *The New York Times*, creates a nightmare: "For the first time, nuclear war might be made thinkable."[20]

In fact, as emphasized by Colin S. Gray, a leading ideologist and proponent of SDI, the United States is posited

> To conduct the Strategic Defense Initiative, in conjunction with modernization of the strategic forces. . . . Mixed offensive-

---

19. Text in *The New York Times*, March 24, 1983.

20. Leslie H. Gelb, "'Star Wars' Advances: The Plan and Reality," *The New York Times*, December 15, 1985.

defensive capability is the historical norm. The real challenge to the U.S. defense community is to plan for a complex environment wherein both superpowers, prospectively permanently, will have offensive and defensive forces in shifting combination.[21]

This determination to proceed with a gigantic effort to conceive and develop new "defensive" weapons systems, and at the same time, as emphasized by President Reagan, "to remain constant in preserving the nuclear deterrent,"[22] i.e., constantly improving and refining the nuclear arsenals, has become a fundamental feature of the SDI drive.

With the current intensification of the arms race, military R&D has experienced an exponential increase in new investments. This added impulse for military technology is full of grave political and military consequences: the arms race is propelled into the future, and dangerous new weapons systems may see the light of the day; there is a growing probability of accidental breakdowns of increasingly complex military constructs. The way military technology will be addressed and handled in the near future is thus of vital importance for us all.

Restraint of Military Technology — The Arms Control Imperative

One main cause for the failure of concerned bodies and the peace movement to moderate the course of the arms race lies in the fact that the evolution and transformation of the arms race from quantity to quality has not been clearly perceived. Consequently, the debate on disarmament and arms control has followed the bounds set by the Establishment—the "numbers game," as dubbed by Alva Myrdal—rather than focusing on the critical issue, which is the unrestrained expansion of military technology. Partly due to the secrecy surrounding military R&D, we have failed to take up the problems related to the emergence of new weapons systems at the moment of their

---

21. Colin S. Gray, "The Transition from Offense to Defense," *The Washington Quarterly* 9, No. 3 (Summer 1986), pp. 66, 71–72.

22. *The New York Times*, March 24, 1983.

inception in the military laboratories; instead, we have tried to intervene, too late, only at the moment of their imminent deployment. As mentioned above, one fundamental lesson in the push and pull dynamics of new weapons systems is that whenever the technological feasibility of new weapons systems is proved in the laboratories, their production and deployment can scarcely be halted.

We urgently need to make the current runaway momentum of the arms race more transparent and to develop strategies to restrain military technology. The race to arm outer space requires preemptive action to curb military technology before it is too late. We may have perhaps a few years' time to decelerate, halt and reverse this mindless arms race. If not, we may find ourselves in an increasingly overarmed world filled with new and deadly weapons, with an ever greater danger to our security and peace.

Obviously, it is a tall order to require the redirection of arms control negotiations from mainly quantitative to a substantially qualitative approach aimed at restraining military technology. However, given political will and concerted efforts, it may not be as intricate as attempts to achieve a quantitative balance in profoundly diverse force structures. Given such a reorientation, negotiators, while perservering in efforts to sharply reduce existing military arsenals, would have to concentrate on controllable and verifiable ways to constrain the engine of the contemporary arms race: military technology.

Reorienting the arms control process is likely to require a sustained long-term effort. We could, however, make a start immediately, by imposing restraints on those stages of military R&D which are observable, controllable and verifiable—such as a comprehensive nuclear test ban and a ban on flight-testing of ballistic missiles. The production of fissionable materials for nuclear weapons can also be subjected to strict control. Moreover, technical measures can be supplemented by restricting funding for military R&D and by establishing national and international technological assessment bodies to act as early warning systems against excesses of military technology. Long overdue is the establishment of a UN International Satellite Monitoring Agency, which could make control and surveillance open for the international community.

Yet technical restraints and control measures may not be enough. The underlying socioeconomic and structural issues must be addressed as well. This calls for well-planned conversion of military R&D for productive purposes, to respond to the requirements of the civilian economy and to satisfy the basic unmet needs of society. Within this framework, scientists and engineers currently employed by military R&D could be offered a better satisfaction of both their ambitious engagement in R&D and their "bread-and-butter" problems. They could also be mobilized to refrain from engagement in military R&D on moral and ethical grounds. Conversion of military R&D for productive purposes should be seen as a major strategy for redeeming science and technology for their real calling, which is not destruction but the betterment of the human condition.

Transcending Technological Determinism

The post-World-War-II rise of modern military technology and the shift of the center of gravity of the arms race from quantity to quality has rendered the arms race far more intricate and intractable. In an arms race which is mainly quantitative, rising costs and rising levels of armaments will tend to inhibit its acceleration. However, today's shift to qualitative technological competition—with a long-haul thrust projected far into the future and the rapid obsolescence of weapons systems—has made no cost seem too high to preempt the presumed technological lead of the adversary.

Thus, the technological momentum acts in various ways to amplify armaments escalation. Nevertheless, we should beware of perceiving this technological drive as a kind of determinism which forces us to swim with the technological tide. The race in military technology is but a cancerous growth on the body of science and technology and on international relations. As such, it requires surgical treatment, not submission. It needs to be transcended and eradicated, replaced by a sound competition and international cooperation in applying modern science and technology for the improvement of the lot of humanity.

The current sharp acceleration of the race in military technology has an ominous significance beyond the arms race and the danger of nuclear war. The fact that the lion's share of scientific-technological human and material resources goes to military purposes has a profoundly evil impact on the human condition, with far-ranging consequences for the future of humanity. At stake is the nature of our scientific-technological enquiry, indeed the very fate of our civilization. Even if in the short run we should be lucky enough to escape nuclear cataclysm, in the long run humanity may still be doomed to spiritual and material ruin. Ultimately, the survival of humanity is at issue.

Science and technology today exert a pervasive impact on society and on our way of life. At the same time, the symbiosis of advanced science and modern technology has increased their perennial double-edged potential for good and evil. We must be on guard against their being misused for destructive purposes and war preparation. Science and technology must adhere to their mission to contribute to the good of the human species.

III
Technology and Economics

7

The Impact of Technological Change on East-West Economic Relations

MICHAEL KASER

"It is indisputable in both theory and in practice that the interest of the working people as masters of production is the strongest incentive and the most powerful driving force for the acceleration of socioeconomic, scientific and technical progress." Mikhail Gorbachev's linkage of technological advance to the system of incentives could not be quoted to the Institute for East-West Security Studies Conference in Helsinki because his speech on a "New Economic Mechanism" came two weeks later, at a Plenum of the Central Committee of the Soviet Communist Party on June 26, 1987. Two years and three months had elapsed since Mr. Gorbachev became General Secretary in March 1985 and had, at a Central Committee Plenum the following month, launched a program aimed at promoting technical progress to reverse a serious decline in Soviet economic growth.

Well before the Institute's Helsinki Conference, nevertheless, the deficiencies of Soviet central planning in applying scientific discoveries and technological innovation were evident. In the USSR and Eastern Europe systemic factors retarded the effective utilization of technology in comparison with the United States, Japan and Western Europe, despite equivalent allocations of money and manpower for research and development. As Ryszard Frelek, head of the Political Strategy Division of the Polish Academy of Social Sciences, stated at the conference, Poland and the Federal Republic of Germany have about the same stock of engineers but very disparate records in innovation.

Two remedies are open to the governments of Eastern Europe and the USSR: one is to increase the use of Western research in Eastern production by obtaining more equipment embodying it or the "know-how" of it; the other is to adapt the management mechanism, particularly with respect to incentives for innovation. Both have their impact on East-West economic relations and involve three related concepts—the "technology gap," the "imports first" policy in relation to comparative advantage, and the "absorption" of transferred technology.

In defining "East-West economic relations," this study follows the definition of the major OECD technology transfer project, to which the present writer was a contributor, *East-West Technology Transfer: The Trade and Economic Aspects*.[1] The term "West" refers to OECD members while "East" refers to the six European members of Council for Mutual Economic Assistance (CMEA): Bulgaria, Czechoslovakia, the GDR, Hungary, Poland and Romania (collectively "Eastern Europe") and the Soviet Union. That project did not cover "technology transfer" in as wide a definition as previous OECD reports had done: "a process whereby innovations (new products or know-how) obtained in one country are then transmitted for use to another."[2] Techniques which are transmitted by sale of publications or by word of mouth (such as by scientific books, by training schemes and by the communication of "know-how") were not pursued by the recent OECD project, which was limited to the transfer of technology embodied in equipment or in technical documentation such as licenses or industrial cooperation agreements. The availability of East-West trade statistics dictated the concentration of this study on "embodied" technology, but, as Mr. Gorbachev's reform indicates, the Eastern countries can improve their economic performance also by

---

1. Helgard Wienert and John Slater, *East-West Technology Transfer: The Trade and Economic Aspects* (Paris: OECD, 1986), pp. 13–14.

2. Eugene Zaleski and Helgard Wienert, *Technology Transfer Between East and West* (Paris: OECD, 1980); Congressional Research Service, *Technology Transfer and Scientific Cooperation Between the United States and the Soviet Union: A Review* (Washington, DC: U.S. Government Printing Office, May 26, 1977).

"disembodied technical progress," as indeed they did after the weakening of the strict controls associated with the Cold War of the late 1940s and much of the 1950s.

Since the 1960s Eastern buyers of equipment and of materials and components manufactured by advanced processes from the West have been motivated either by the requirements of planned economic development or by priced comparisons of the imported against the domestic product when the exchange rate of the buyer's currency reflects the purchasing power of its exports. The first criterion was formalized into a procedure termed "imports first."[3] A set of requirements offset against availabilities from domestic suppliers defines the desired imports, the sum of which in likely foreign prices is compared with the anticipated aggregate of exports (determined by planned surpluses of availabilities over home requirements). In so simple a form the policy of planning imports first was never applied without some regard for comparative costs,[4] and the sphere of cost-based evaluations was enlarged as enterprises become more closely involved in foreign-trade transactions.[5] A more advanced technology embodied in a piece of foreign equipment must be evaluated in terms of its enhanced performance over the corresponding domestic plant and the difference, suitably discounted over the expected operational life, compared with the value forgone in the exports which would pay for it. That differential will be larger the wider the "technological gap" between the purchasing and the supplying country.

---

3. Described in Philip Hanson and Michael Kaser, "Soviet Economic Relations with Western Europe" in Richard Pipes, ed., *Soviet Strategy in Europe* (New York: Crane, Russak [for Standard Research Institute], 1976), pp. 213–217.

4. The "Methodological Instructions" of the USSR State Planning Commission covering the 1976–1980 and then the 1981–1985 five-year plans came close to a disregard for comparative costs in the reverse case, namely that equipment should not be imported if it could be made at home.

5. The history of such decentralization of foreign-trade decision-making is well set out in chapters by Harriet Matejka in Hans-Hermann Hohmann, Michael C. Kaser and Karl C. Thalheim, eds., *The New Economic Systems of Eastern Europe* (London: Hurst, 1975), pp. 443–477; and in Michael C. Kaser, ed., *The Economic History of Eastern Europe, 1919–1975*, vol. III. (Oxford: Oxford University Press, 1986), pp. 250–288.

Machinery Trade and the Technology Gap

Two tables compiled by the secretariats of the United Nations Economic Commission for Europe and of the OECD display the substantial eastward flow of equipment for investment over the past quarter-century.[6] Table 1 shows the value by five-year period of such sales: during 1961-1985 the USSR bought $102 billion of Western equipment and Eastern Europe $87 billion, all values converted to dollars of 1975 purchasing power. Each flow represented about one-third of total imports of equipment, for the USSR bought from Eastern Europe some $205 billion of machinery in 1961-1985 and the nations of Eastern Europe bought from each other and from the USSR some $211 billion. In most cases the heaviest importation took place in 1976-1980 when Western banks were lending virtually without restriction to Eastern Europe. The measure of this in Table 1 is the "vintage," as economists conventionally term a year of introducing capital equipment into use. For Czechoslovakia, the GDR, Hungary and Romania between one-quarter and one-third of the entire twenty-five-year purchases was concentrated in those five years; for Poland nearly two-fifths were imported in that quinquennium. All those countries slackened their buying after 1981, when Western lending abruptly halted as Poland and Romania failed to service their outstanding debt. Bulgaria, which, like Czechoslovakia, had pursued a cautious policy on convertible-currency indebtedness, made its biggest run of purchases in 1981-1985. Those five years were also the peak of machinery imports from the West for the USSR: 38 percent of the imports of 1961-1985 were effected then in consequence of the USSR's enhanced convertible-currency purchasing power when the OPEC oil price was at its highest ever—$32 per barrel between late 1979 and early 1983 and $28 between then and late 1985.

Table 2 indicates the net balances of trade by commodity group for the key period of 1971-1983. The deduction of Soviet exports of machinery (capital goods and intermediate goods contributing to technology) from its imports of those goods in the twelve-year period shows a net Western sale of $64 billion;

---

6. All recalculated to 1975 prices and dollar values by the UN Economic Commission for Europe Secretariat in "Eastern Imports of Machinery and Equipment, 1960-1985," *Economic Bulletin for Europe* 38, No. 4 (December 1986), Table 3.5.

Table 1
Eastern Europe and the Soviet Union: Imports of machinery, five-year cumulations and vintage structure, 1961-1985
(In billion of 1975 U.S. dollars and percent)*

	1961-1965	1966-1970	1971-1975	1976-1980	1981-1985	1961-1985
From non-socialist countries						
Bulgaria	0.41	1.26	1.56	1.70	2.92	7.85
	(5.2)	(16.1)	(19.9)	(21.6)	(37.2)	(100.0)
Czechoslovakia	0.83	2.41	3.79	4.48	4.15	15.66
	(5.3)	(15.4)	(24.2)	(28.6)	(26.5)	(100.0)
German Dem. Republic	0.80	2.52	4.46	5.01	5.97	18.75
	(4.2)	(13.4)	(23.8)	(26.7)	(31.9)	(100.0)
Hungary	0.42	0.80	1.57	2.31	1.99	7.09
	(5.9)	(11.3)	(22.1)	(32.6)	(28.1)	(100.0)
Poland	1.07	2.39	7.88	8.89	3.20	23.43
	(4.6)	(10.2)	(33.6)	(38.0)	(13.6)	(100.0)
Romania	1.11	2.59	4.13	4.38	1.71	13.91
	(7.9)	(18.6)	(29.7)	(31.5)	(12.3)	(100.0)
Eastern Europe	4.64	11.96	23.38	26.77	19.95	86.70
	(5.3)	(13.8)	(27.0)	(30.9)	(23.0)	(100.0)
Soviet Union	6.26	10.78	17.37	30.45	37.52	102.38
	(6.1)	(10.5)	(17.0)	(29.7)	(36.6)	(100.0)
From socialist countries						
Bulgaria	2.57	4.27	6.86	8.76	10.99	33.46
	(7.7)	(12.8)	(20.5)	(26.2)	(32.8)	(100.0)
Czechoslovakia	3.64	5.10	9.13	13.20	14.76	45.84
	(7.9)	(11.1)	(19.9)	(28.8)	(32.2)	(100.0)
German Dem. Republic	2.05	5.13	9.98	13.14	12.47	42.77
	(4.8)	(12.0)	(23.3)	(30.7)	(29.2)	(100.0)
Hungary	1.51	2.20	4.15	6.20	6.17	20.24
	(7.5)	(10.9)	(20.5)	(30.7)	(30.5)	(100.0)
Poland	3.65	5.78	10.08	14.10	12.40	46.01
	(7.9)	(12.6)	(21.9)	(30.6)	(27.0)	(100.0)
Romania	1.97	2.60	4.42	6.72	6.49	22.20
	(8.9)	(11.7)	(19.9)	(30.3)	(29.2)	(100.0)
Eastern Europe	15.40	25.08	44.62	62.13	63.28	210.51
	(7.3)	(11.9)	(21.2)	(29.5)	(30.1)	(100.0)
Soviet Union	14.98	20.96	35.29	55.59	78.01	204.82
	(7.3)	(10.2)	(17.2)	(27.1)	(38.1)	(100.0)

Source: ECE secretariat estimates, in *Economic Bulletin for Europe* 38, No. 4 (December 1986), p. 618.
*Vintage structure is in parenthesis.

Table 2
Composition of Cumulative OECD Surpluses on East-West Trade by Selected Commodity Groups 1971-1983
(In billion of current U.S. dollars)

	1971-1975	1976-1980	Total	1981-1983
Soviet Union				
Capital goods	10.5	29.0	39.5	19.8
Intermediate (technology) goods	7.0	17.4	24.4	13.5
Fuel	−10.9	−44.2	−55.1	−52.0
Other industrial goods	−2.3	−4.1	−6.5	3.9
Total industrial goods	4.3	−1.9	2.4	−14.7
Agricultural goods	0.9	7.4	8.3	9.1
Total	5.2	5.5	10.7	−5.7
Eastern Europe				
Capital goods	10.7	17.4	28.1	6.2
Intermediate (technology) goods	9.3	15.0	24.3	4.2
Fuel	−4.4	−12.3	−16.7	−9.8
Other industrial goods	−5.3	−9.8	−15.1	−4.9
Total industrial goods	10.3	10.3	20.6	−4.3
Agricultural goods	−0.1	5.5	5.4	3.0
Total	10.2	15.8	26.0	−1.2
Total, Seven Eastern Countries				
Capital goods	21.2	46.4	67.6	26.1
Intermediate (technology) goods	16.3	32.5	48.7	17.7
Fuel	−15.2	−56.5	−71.8	−61.8
Other industrial goods	−7.6	−14.0	−21.6	−1.0
Total industrial goods	14.6	8.4	23.0	−19.0
Agricultural goods	0.8	12.9	13.7	12.1
Total	15.4	21.3	36.7	−6.9

Note: Symbol "−" = Western deficit. The Table is based on Western exports *f.o.b.* and Western imports *c.i.f.*; hence Western surpluses tend to be understated and deficits overstated.

Source: OECD East-West technology transfer data base (Western statistics) in H. Wienert and J. Slater, *East-West Technology Transfer: The Trade and Economic Aspects* (Paris: OECD, 1986), p. 228.

for Eastern Europe the net Western sale was $52 billion. Soviet payments in oil and gas are shown by the net Soviet sale of fuel of $55 billion, and Eastern Europe counter-deliveries by $17 billion of fuel (much of it Polish coal and other countries' refined products of Soviet oil) and $15 billion of current manufactures, mostly consumer goods. Taken together, Eastern Europe and the USSR accumulated over the period of 1971 to 1983 a net visible deficit of $37 billion with OECD members.

Important as these supplies were in abating the Eastern "technology gap," they never ranked highly in Western sales. Over the entire twenty-five years covered by the ECE data, the share of Eastern Europe and the USSR in Western exports varied between 3.2 percent in the early 1960s to 4.3 percent in the late 1970s and the counter-deliveries from 3.25 and 3.6 in the same subperiods.

A major study by the Centre for Russian and East European Studies of the University of Birmingham some ten years ago demonstrated the Soviet lag behind advanced industrial Western countries at all stages of the research-production cycle. Robert Davies, then director of the Centre, summarized their conclusion that "in most of the technologies we have studied there is no evidence of a substantial diminution of the technological gap between the USSR and the West in the past fifteen to twenty years, either at the prototype/commercial application stages or in the diffusion of advanced technology."[7]

The same Centre has found evidence that the technology gap has subsequently widened. Presenting various Soviet time-series of indicators of technological progress, Ronald Amann, the Centre's current director, observes that they

> cover the whole of the latter part of the research-production cycle: invention, innovation, diffusion and incremental improvement. Compared with achievements in scientific research, these are the phases of development where it is generally thought that the USSR exhibits relative weakness as a result of discontinuities between different phases of development and lack of incentives.... The general picture that emerges from the statistics is extraordinarily striking and consistent. There has been a continuous absolute decline since the early 1960s in the creation of prototypes of new machines and equipment, espe-

---

7. Ronald Amann, Julian Cooper and R.W. Davies, eds., *The Technological Level of Soviet Industry* (New Haven, Connecticut: Yale University Press, 1977), p. 66.

cially in the more advanced science-based industries.... The next stage in the research-production cycle is the introduction of the new technologies into the economy and their economic impact. Once again one can see a fall in the rate of growth of these measures since the first half of the 1970s.... The most startling trend to emerge ... is the relatively steep decline of the impact indicators, expressed in terms of labor savings from new technologies and their annual economic effect. This suggests a growing ineffectiveness of industrial innovation.... [I]ncremental improvements at the shop floor level ... are ... an important aspect of technical progress in the West. This is where science interacts with production. In the course of fully mastering and modifying a new process fundamentally new technologies can begin to take shape. In the USSR, however, the long-term statistical trends would seem to cast some doubt on the vitality of these creative responses. Since the early 1970s there has been a fall in the growth rate of the number of improvements introduced and a similar though less pronounced fall in expenditures on their implementation.[8]

The two studies on the industrially and technologically advanced countries of Eastern Europe, Czechoslovakia and the GDR, induce similar conclusions. Friedrich Levcik and Jiri Skolka observe in a major study for the OECD project that "Czechoslovak data show a deterioration in the technical level of Czechoslovak production during the 1970s.... The number of [new products] classified as being at 'world technical and economic level' halved between 1970 and 1977."[9] An independent analysis by Raymond Bentley drew the conclusion that "the German Democratic Republic appears to have had an especially pronounced [technological] lag behind the Federal Republic of Germany in the more modern branches of technology such as electronics, data processing, instrument building, synthetic fibers and plastics. The lag is probably due more to endogenous than exogenous factors."[10]

None of the studies cited belittle the very considerable

---

8. "Introduction" in Ronald Amann and Julian Cooper, eds., *Technical Progress and Soviet Economic Development* (Oxford: Blackwell, 1986), pp. 9–10.

9. Friedrich Levcik and Jiri Skolka, *East-West Technology Transfer Study of Czechoslovakia* (Paris: OECD, 1984), pp. 68–71.

10. Raymond Bentley, *Technology Change in the German Democratic Republic* (Boulder, Colorado: Westview Press, 1984), pp. 174 and 226.

Soviet and East European achievements in scientific research, above all in the past thirty years (a Nobel Prize for science was first awarded to a Soviet citizen in 1956). Progress would have been faster over the previous period of the Soviet state, when "pseudo-scientific approaches became dominant, or at least prominent, in soil science, silviculture, zoology, botany, evolution, agrochemistry and many other areas."[11] The substantial employment in, and expenditure on, scientific research in all the Eastern European countries have yielded their returns, though on a world scale not proportionate to those inputs. Thus in 1980 Soviet employment in science, research and development (1.4 million) exceeded the aggregate similarly employed in the United States, Japan, the FRG, the United Kingdom and France put together (1.3 million),[12] but those five Western countries generated more discoveries, however they might be measured, than the USSR.

Furthermore, the application of scientific and technological research results in sales to the West by the USSR and other Eastern European countries, as John Kiser has shown.[13] However, these sales represent a much smaller flow than those in the opposite direction.

Military Allocations and the Technological Gap

It is impossible to escape the conclusion that the East-West technology gap must have a systemic cause, that it arises because the incentives to engage in, and be rewarded by, the research-production cycle differ in a market system and in a planned economy. But the group of systemic inhibitions are themselves entirely responsible for the gap because of the heavy involvement of scientific and technological personnel and resources in military applications.

---

11. Zhores A. Medvedev, *Soviet Science* (Oxford: Oxford University Press, 1979), p. 55.

12. Yevgeni Lazarev, "International Exchange of Scientific and Technical Knowledge," *USSR Foreign Trade*, No. 2, 1987, p. 37, citing *Science Indicators* (Washington, DC: National Science Board, 1985).

13. John W. Kiser, *Commercial Technology Transfer from Eastern Europe to the United States* (Washington, DC: United States Department of Commerce, 1980).

Before establishing this argument in detail, the context of the defense sector within the economy must be delineated. That context primarily concerns the Soviet Union, where estimates of the share of gross national product (GNP) devoted to defense range (for 1980) from 8.8 percent by the Stockholm International Peace Research Institute to 13–14 percent by the U.S. Central Intelligence Agency and 15 percent by the U.S. Defense Intelligence Agency. Two American researchers consider their government's estimates too low—for 1975 when the official agencies were close to each other at 13–14 percent of GNP, one had estimated 15.3 and the other 15.5 percent.[14]

The other East European economies devote a much lower share of their resources to defense; the total percentage is more readily found in their own calculations of net material product (NMP) which falls short of GNP by most services. Because military personnel (as well as those in health, education, government and finance) are excluded from NMP, the percentage is not "of" national income, but "compared against" national income. The mean of the CIA and DIA estimates for the USSR in 1980 (83.5 billion rubles) represents 18 percent when set against NMP, compared with 5 percent in Bulgaria, Czechoslovakia, Hungary and Poland and 7 percent in the GDR.[15]

Noting that "even under current spending programs of the Reagan Administration, the U.S. ratio is not likely to rise above 8–9 percent" and "assuming that something like 13–15 percent is an acceptable reading of the current ratio of Soviet military outlays to GNP," Abraham Becker concludes that "the measured defense/GNP ratio understates the actual proportion of Soviet aggregate output that is devoted to military-security requirements. This is probably also true in the West, but the gap is surely larger in the USSR." He argues, as have others,

---

14. Tabulated by Abraham S. Becker, *Sitting on Bayonets. The Soviet Defense Burden and the Slowdown of Soviet Defense Spending* (Los Angeles, California: RAND/UCLA Center for the Study of Soviet International Behavior, 1985), p. 13. The sources are there listed, the independent estimators being William T. Lee, *The Estimation of Soviet Defense Expenditures, 1955–1975* (New York: Praeger, 1977), and Steven Rosefielde, *False Science: Underestimating the Soviet Arms Build-Up* (New Brunswick, New Jersey: Transaction Books, 1982).

15. Romanian defense expenditure is not published in any usable form; the official data for the other countries have been adjusted to approximate comparability to the estimates for the USSR.

that

> military industry is supplied with scarce, high quality resources often unavailable to civilian industry; the pick of production in dual-line plants may be taken for military needs, leaving the inferior product for civil use; in the event of shortages, military programs tend to be protected, leaving civil activities to cope as they can. In addition, the walls of insulation that for so long sheltered the military economy and that still today are only partly permeated have hindered spillovers of usable military innovation in products, processes or, to a lesser extent, organization.[16]

Estimates of the share of research and development in the USSR allocated to military use range from one-half to two-thirds.[17] Julian Cooper warns, however, that

> Western writings on the Soviet defense industry have tended to portray it as a separate distinct sector walled off from the civilian economy by almost impenetrable barriers of secrecy. With this image has been associated the view that transfers of technology from the defense industry are almost nonexistent. In the author's opinion, reinforced by the present research, these traditional perceptions require modification: the boundaries between the sectors are more fluid and the transfers more extensive than is generally believed.[18]

While allowance must be made for spinoffs to civilian uses, the cost of military research and development is considerable in the field of equipment and materials, many of which embody advanced technology. The present writer made a highly tentative estimate that it absorbed 17 percent of all expenditure on physical goods in 1979, the other absorptions of goods (as distinct from personnel) being 46 percent to procurement of materiel and 37 percent to investment.[19] Mr. Gorba-

---

16. Becker, "Sitting on Bayonets," p. 21.

17. Evidence of Ronald Amann and Philip Hanson, in House of Commons, Second Report from the Foreign Affairs Committee, Session 1985–86, *UK-Soviet Relations*, Volume II (London: Her Majesty's Stationery Office, 1986), p. 230.

18. Cooper in Amann and Cooper, p. 48.

19. Kaser in Archie Brown and Michael Kaser, eds., *Soviet Policy for the 1980s* (Bloomington, IN: Indiana University Press, 1982), p. 206.

chev is reported to have said to President Reagan during the informal meetings at the Geneva summit in November 1985 that he wanted "to transfer billions of rubles from the defense to the civilian budget"[20] and there can be no doubt that such a release would accelerate Soviet technical progress. It is by confining technological resources to military uses that the "defense burden" is particularly impeding. This is not to say that the Soviet government under Mr. Gorbachev has been attracted towards arms control negotiations by an inability to match United States or NATO defense expenditure. It does mean that his policy of "acceleration" would benefit substantially by the transfer of research and development outlay—benefitting the civilian economy by more than it reduces military output. The gain from transfer arises because the higher quality and technology of defense resources would yield greater returns than the resources currently in civilian use. In fact, it is possible to show that the quality of resources earmarked for the military improved during the 1970s when Mr. Brezhnev's administration was achieving parity in strategic weapons systems and deployments with the United States. The rise in efficiency meant that the USSR was "getting more bang for the buck" in the colloquial phrase, for the military-industrial sector became more cost-effective while the civilian-industrial sectors became less cost-effective.[21]

By the same token, any future transfer out of the civilian research and development to the military would impose a marginally much heavier burden. It is surely for this reason that Mr. Gorbachev went so far at the Reykjavik summit to link his acceptance of NATO's "zero option" on intermediate-range nuclear forces (INF) in Europe to a required United States deferment of experimental deployment of the Strategic Defense Initiative (SDI) outside laboratories. Matching SDI beyond existing anti-ballistic missile (ABM) preparations would make a serious hole in civilian economic development.

Systemic Factors in the Technology Gap

The Soviet civilian economy as it was operating in the 1970s under Brezhnev after the demolition of Kosygin's moderate

---

20. Unconfirmed contemporary press reports.

21. Kaser in Brown and Kaser, p. 222.

economic reform had many built-in hindrances to efficient exploitation of the research-production cycle. Two weighty studies, by Amann and Cooper and by Joseph Berliner, provide the case studies and draw useful conclusions regarding the system's shortcomings in comparison with a market economy.[22]

Innovation involves risk that the application of the technology may not conform to expected outcomes. Planners and executants alike are averse to uncertainty and risk and hence have reservations on innovation. The objective of both planners and enterprises has been in the past the maximization of production within a defined period. That is often sought "at the expense of the quality and novelty of products" and hence in a process conducive to "a relatively low priority to industrial innovation."[23] In the Soviet administrative structure, technological research and development is isolated from production facilities; each set of the latter under one industrial ministry is insufficiently linked to those of another ministry. Incentives have been weak for research and development agencies. "The complexity and artificiality of salary and bonus calculations in the enterprise sector extends also to the specialized industrial R&D sector, giving rise to unintended distortions."[24] There is a disjunction between production and specialized research institutions (very little applied research takes place in Soviet universities, universities which until Mr. Gorbachev began a reform were "teaching factories" rather than research agencies). The pricing mechanism often acts as a disincentive to innovate.[25] Among the consequences is a slowness, even inertia, in applying proven innovations. Two United States observers have found that after two years only 23 percent of patents

---

22. Ronald Amann and Julian M. Cooper, *Industrial Innovation in the Soviet Union* (New Haven, Conn.: Yale University Press, 1982), especially pp. 11–30; Joseph S. Berliner, *The Innovation Decision in Soviet Industry* (Cambridge, Mass.: The MIT Press, 1976), especially pp. 475–541. For a more general review, see Jozef Wilczynski, *Profit, Risk and Incentives under Socialist Economic Planning* (London: Macmillan, 1973).

23. Cooper in Amann and Cooper (1986), p. 16.

24. Amann in Amann and Cooper (1982), p. 14.

25. Examined in great detail by Berliner, *The Innovation Decision*, pp. 235–396.

(known in the USSR as "certificates of authorship") had been introduced into production; in the same time period 66 percent had been applied in the U.S. and 64 percent in the FRG.[26]

Technology Transfer from the West

Two major, if conflicting, accounts are available of the importance to the USSR's economic development of Western technology. Anthony Sutton in a trilogy covering fifty years of numerous case studies contends that Western technology greatly contributed to Soviet development, while Philip Hanson for the period 1955 to 1979 shows imports repriced into domestic investment costs rising from 2 to 6 percent by 1975 and falling to 4 percent.[27]

For the Eastern European countries the concept of "import-led growth," initiated by Stanislaw Gomulka,[28] has had considerable influence. As summarized in the ECE Secretariat study, "according to this view there has been a growing perception in the East of a widening technological gap, with respect to the most industrialized market economies. In an attempt to improve their position, the Eastern countries thus stepped up their imports of advanced technologies considerably in the early 1970s."[29] Payment for such imported equipment, the

---

26. John A. Martens and John P. Young, "Soviet Implementation of Domestic Inventions: First Results," in U.S. Congress, Joint Economic Committee, *Soviet Economy in a Time of Change* (Washington, DC: U.S. Government Printing Office, 1979), pp. 472–509.

27. Anthony Sutton, *Western Technology and Soviet Economic Development*, 3 vols. (Stanford, California: Hoover Institute Press, 1968, 1971 and 1973); Philip Hanson, *Trade and Technology in Soviet-Western Relations* (London: Macmillan, 1981); and in Brown and Kaser, especially Table 3.9, p. 80. See also Bruce Parrott, ed., *Trade, Technology and Soviet-American Relations* (Bloomington, Ind.: Indiana University Press, 1985).

28. Stanislaw Gomulka and J. Sylwestrowicz, "Import-led Growth. Theory and Estimation" in Franz-Lothar Altmann *et al.*, eds., *On the Measurement of Factor Productivities* (Goettingen: Vandenhoeck and Ruprecht, 1976), pp. 539–574; and Stanislaw Gomulka, "Growth and the Import of Technology: Poland 1971–1980," *Cambridge Journal of Economics* 2 (1978), pp. 1–16; and Philip Hanson, "The End of Import-Led Growth," *Journal of Comparative Economics* 6 (1982), pp. 130–147.

29. *Economic Bulletin for Europe*, December 1986, p. 59.

explanation goes, was financed by extensive borrowing in the West (which did, of course, take place until the rescheduling crisis of 1981-1982), intended to be repaid by the export goods engendered by the reequipment of domestic industry. That this strategy failed is partly attributable to the downturn in Western economic activity and hence to the East's export markets.

The ECE Secretariat has shown, however, that the import-led strategy is valid only for Poland, the USSR "and perhaps Bulgaria."[30] That it was not inspired by the availability of "easy" Western money (the "petrodollars" which rendered Western banks very liquid from late 1973 onward) is evidenced by the start of the equipment import surge in the 1960s. Moreover, the borrowed funds predominantly purchased current, rather than capital, goods. The ECE Secretariat's findings are that "only Poland during 1972-1975 and Czechoslovakia in 1976-1978 appeared to have devoted a relatively high share of their new credits to finance machinery imports for any significant period of time. In general, however, Eastern borrowing does not seem to have been associated with raising machinery imports."[31]

The OECD study concludes that in Eastern Europe "in ranking import priorities ... greatest emphasis has been placed on feeding the population; maintaining supplies of industrial inputs, especially intermediate (technology) goods, has taken second place and that supplying new technology in the form of imported capital goods has ranked third."[32]

Western technology has contributed to Eastern economic development but has not played a dominant role.

Disembodied Technical Progress

Reference must be made in conclusion to the reform which Mr. Gorbachev is using to complement and partly substitute for technology embodied in capital goods. He is dismantling economic Stalinism just as Khrushchev exactly thirty years earlier, at the Twentieth Party Congress in February 1956, broke the grip of political Stalinism. The disappearance of many of the

---

30. Ibid., p. 80.

31. Ibid., p. 81.

32. Wienert and Slater, *East-West Technology Transfer*, p. 201.

props of the "command economy" (which Stalin imposed so that no orders should prevail save those he gave or delegated) will reveal some of the structure that was there under NEP—notably peasant agriculture and small handicrafts—but a new economic mechanism is having to be created while the old is withdrawn. It is illuminating that the term Gorbachev uses for the process he has now signally intensified, *perestroika*, has not only the sense of "reconstruction" or "restructuring" but also of "transformation." On one occasion Gorbachev (in a speech at Khabarovsk in July 1986) went so far as to observe, "'Restructuring' is a very meaningful word. I would put an equals sign between the words 'restructuring' and 'revolution'."

But it is not a "revolution" he is seeking—though it is such for the officials sacked for incompetence or inadaptability or jailed for corruption—because he could not tolerate the disturbance inevitable in any fundamental change. It is not simply that he is powerless to displace existing bureaucracies: he wants their support, and their efficient work, for the objective of "restructuring" is to keep the party through which he rose in permanent (though no longer absolute) power. Rather, he has linked "restructuring" with "acceleration": when more outputs are to be obtained from limited inputs (a manpower deficit, the arms burden and pressing claims on investment), it is not time for "productive forces" to lie idle while "productive relations" are regrouped.

At the Central Committee Plenum of June 1987 he made enterprise autonomy the crux of an efficient "New Economic Mechanism":

> The main question in the theory and practice of socialism is how, on a socialist basis, to create more powerful stimuli than under capitalism, for economic, scientific, technical and social progress, how to most efficiently combine planned leadership with the interests of the individual and of the collective. This is a most complex question the answer to which has been sought and is being sought by socialist thought and by social practice. At the present stage of socialism the significance of this question is increasing immeasurably.

But he immediately countered the argument that the revolution of central management was "virtually the restoration of private economic practice":

> I think, comrades, that our new experience and that of the other

socialist countries attests to the usefulness of and need for the skillful exploitation of these economic forms within the socialist framework. They help to satisfy people's urgent needs more fully and remove the "black" economy and all possible forms of abuse, that is, they help the real process of improving socioeconomic relations.

There also needs to be a serious rethinking of the problems connected with the relationship between centralized planned leadership of the national economy and the independence of its individual components, planning and commodity money relations. We proceed on the basis of their dialectical unity and complementarity in an integral system of economic management.[33]

It was the Stalinist command economy which supported the political autocracy—and united both into a form of totalitarianism—and it is a plurality of currents in both which Gorbachev promises under concepts of "restructuring" and "democratization." Totalitarian rule at home and expansionism abroad were then inevitably met by opposition on the part of the Western democracies, but as they see those practices weaken it becomes their interest to help them wane more rapidly. Reversal being to their disadvantage, governments and business should not take measures which would afforce the considerable domestic opposition to Gorbachev.

---

33. Speech by Mikhail Gorbachev to the June Plenum of the Party Central Committee, *Pravda*, June 26, 1987.

8

Economic Forces in the History of East-West Relations

JOZSEF BOGNAR

Introduction

The technological revolution, as well as the economic changes linked to it,[1] has been the most important international development since World War II. It has had a powerful influence not only on various countries and regions but also on the belief systems and vested interests which influence the way these countries and regions shape their economic and political priorities. On a global level, it has introduced new trends in relations among states and institutions.

The acceleration in technological progress has intensified contradictions between technology, the environment (or biosphere) and society, including international relations. The increased danger to the biosphere is now a well-understood phenomenon, one which calls for the destructive effects of present-day technology to be reduced and new development tasks to be solved by technologies that are not harmful to the biosphere. The intensification of contradictions between technology (the technosphere) and the biosphere has important economic, legal and international consequences. First, goods have ceased to be "free" in the economic sense, because the utilization of air and water have become cost factors. Second, from a legal viewpoint, the environmental effects of economic

---

1. Jozsef Bognar, *Vilaggazdasagi korszakvaltas* (New Era in the World Economy) (Budapest: 1976).

growth represent an international, not merely a national, problem. Finally, the solution of important environmental problems requires international agreements, the coordination of national budgets, and the introduction of international norms and verification systems.

Accelerated technical progress represents a new challenge to the social sphere as well, in every country and in every kind of social system. Social structures must be sensitive enough to foresee new changes and transform the targets of economic development if the old ones have become outdated. They must be able to adjust the speed of reaction to external changes as well as the pace of liberating internal sources of innovation. In addition, they must take into consideration the technological and economic problems caused by the shortening of the life cycle of products (the "technical rent" now is a lease of only two to three years) while strengthening their capacity for flexible transformation.

These national challenges aside, both the hazards of technology and the strengthening of interdependence must be seen in an international context. As far as the hazards of technology are concerned, one must consider not only military technologies or biology by themselves, but also the complex relationship of direct and indirect dangers which causes John von Neumann to ask "whether we can survive technology"[2]—i.e., whether we have the capacity, flexibility, wisdom and institutional organization to protect us from catastrophe.

Conversely, interdependence means that we are dependent on each other not only in military-security terms—and this includes the leading powers—but also in terms of a large number of problems connected with the biosphere and the technosphere, with economic progress—in short, with the progress of mankind in general.

Another challenge we face is due to the fact that in the new era of the world economy, the traditional interconnections between the different production processes established during the period of mass production that made interlinked processes possible have been disrupted. Such linkages include those between production and the energy needed for it; production and the quantity and nature of materials necessary for it; pro-

---

2. John von Neumann, "Can We Survive Technology?" *Fortune*, June 1955.

duction and transportation; the demand for machinery in new and old technologies; the flow of capital and the order of magnitude of trade; production and employment; and the economy and the state budget. The disruption of these interconnections means that less energy and raw materials, as well as less transportation, are needed for the production of the same value today, compared with those production requirements in earlier times. Second, less manpower is needed for modern production. Since the technical revolution in tertiary services is advancing at great speed and has not yet reached its peak, there are presently no "absorbing" trades.

In the past, the flow of capital preceded or followed trade and boomed or receded in close connection with it. Today, however, the flow of money amounts to thirty times that of the flow of goods (i.e., of international trade), signifying that money is being moved under the influence of other factors.

The organic links between economic growth and the state budget also have become disrupted, a fact which contributes significantly to current budgetary deficits all over the world.

The disruption of the interconnections mentioned above means, therefore, that the supply of energy, raw materials and transport capacities is higher than the demand for them, and consequently that their prices are as low as they were prior to the great crisis of 1929. This also means that the export income of the developing countries has fallen, while their accumulation of debt has risen to a level unimaginable in an earlier day. It has thus become obvious that the *accumulation of debts*—which in present circumstances represents the main obstacle to the organic growth of world trade—can be solved only within the framework of a rejuvenated international monetary system.

Finally, it must be taken into consideration that the world economy cannot forever grow extensively. Existing problems cannot be solved by the extension of development to areas which were *terra incognita* earlier, since with the rise of the Third World economic processes cover the entire globe. Consequently, the problems, contradictions and dangers connected with the technical revolution can be resolved only by better cooperation. Following this brief sketch of the new trends and processes in the world economy, I shall now examine the problems connected with them which have arisen in the European socialist countries and in the People's Republic of China.

The 1960s and 1970s

In the second half of the 1960s, the European socialist countries experienced a weakening of the economic forces which had driven their societies during the postwar social transformation and which had stimulated growth despite contradictions within their national economies. As a consequence of this diminished growth, economists and politicians from a variety of backgrounds came to recognize that fundamental economic changes were necessary. Furthermore, the manpower reserves that had become available in the course of the socialist reorganization of agriculture were exhausted, and a chronic shortage of labor occurred in numerous trades.

With the benefit of experience, one can argue that at the time too much capital, material and labor was put into one unit of increment of the national income, and this caused periodic imbalances in the economy. In-depth analysis of these phenomena leads one to the conclusion that the economic model of the postwar period of social transformation is inadequate for solving problems of intensive economic growth, and consequently that reform is needed. Since the mid-1960s the expression "reform" has had two meanings. The more radical economists conceived of reform as covering all economic activity and consequently spreading at an appropriate time to the social-political area, while the more cautious ones spoke only about reform of the system of control and guidance, which meant that the economy should be led by economic levers and not planning instructions.

According to the radical reformers, it was not simply that planning instructions caused the problem, but also that individual and group interests were not integrated into economic activity; commodity and monetary relations were neglected or considered secondary; monopolistic organizations had been established; production and distribution were rigidly separated from each other; and companies had no freedom of action. Particularly grave mistakes were committed in agricultural policy, and, as a consequence, most European socialist countries as well as China became dependent on farm imports. Since choices have to be made concerning food versus technology imports, agricultural imports have a crucial effect on every possible model of economic growth, always leading to a slowdown of industrial development.

External economic components of the model of the period of social transformation are also very important. European socialist countries suffered numerous blockades and embargos that made it difficult, even impossible, for them to adjust the external relations of their economies to the requirements of economic rationality.

It is clear that a planned economy functioning under an embargo is inclined to turn inward, since, for example, it is unable to plan convertible currency exports realistically and sales prices depend on circumstances which are impossible to predict. Consequently, the European socialist countries adopted the idea of the *inevitability of the exchange of goods*, which in practice means that the national economy should produce everything that it can, irrespective of cost and reason, importing only the natural resources which it does not possess or is as yet unable to produce. It is obvious that within such an economic framework external relations become minimized and their forms impoverished. In the period under review, unfortunately, such a situation did not run counter to the intentions of the West vis-a-vis Eastern Europe.

In this context a very modest reform was introduced in the Soviet Union in 1965, usually known as the "Kosygin reform." Every reform is judged by the efficiency (or survival) of each component reform effort, as well as how the quantity and quality of changes being introduced relate to the elements which remain unchanged. It can be stated unequivocally that the quantity and quality of the intended Soviet changes in 1965 were too small to survive in a conservative structure which preferred the status quo.

In 1968, however, an economic reform was introduced in the Hungarian economy, and this reform has survived despite the fact that the tide of reform turned in the East after the well-known events in Czechoslovakia. (In socialist societies, strict, rigid and preponderantly ideological cycles tend to interchange with liberal cycles. A single-party system relying on a broad base of alliances makes such rotation possible by enabling the representation of different political platforms and by supporting representation of different interests through a system of reconciliation.)

The epochal change in the world economy heralded by the energy crisis in the early 1970s brought about a new situation in the socialist economies. The new developments did not affect all the European socialist economies simultaneously, since the first wave of changes (high prices for energy, raw materials and gold) to some degree worked to the benefit of the Soviet economy. They did, however, have an immediate negative effect on countries short of energy and raw materials.

The second wave of world economic changes—a slowdown of economic growth; the stagnation and then only slow growth of world trade; changes in the terms of trade; the credit crisis; high interest rates; and then the sudden acceleration of technical progress—was unfavorable for all socialist economies. Under the influence of these changes, export income fell throughout the socialist community; this was determined structurally, not by a fall in volume. Prices of agricultural products also fell by 40 percent in comparison to 1975. The various socialist countries were therefore unable to reduce imports to any significant extent, resulting in balance-of-trade deficits. They tried to counteract this by raising credits, which was possible until the end of the 1970s. Owing to the credit crisis, however, obtaining loans became more difficult, interest rates rose vigorously, and concurrent debt-service obligations increased.

Debates evolved in the socialist countries about the nature of the world economic crisis. The reformists held to the view that the changes were the consequences of a huge structural change and reflected a new situation in the world economy. Consequently, not only were the changes here to stay; they would also spread to areas still untouched by change. Economic reform must therefore be continued (or undertaken), they argued, but adjustments must be made in order to respond to international developments.

Others, mostly the more conservative economists, voiced the opinion that perhaps a larger than usual business-cycle crisis was occurring, and in the course of this cycle the recession would be followed by a recovery. This meant that the crisis had actually been "postponed." Such a "postponement" had negative consequences. Because the economy had already used up a considerable part of its energies during the primary crisis period, a more serious form of the crisis would have to be confronted later with fewer energy reserves.

The world economic crisis and the increase in tensions gave rise to new systems and efforts in various national economies, leading in the 1980s to a new reform period in the socialist countries.

The 1980s: The Current Tasks of Reform

Beginning in 1979, the economic reform in Hungary (which had been unable to introduce its planned external economic reform owing to the "anti-cycle" which had lasted since 1972) again gathered speed and spread to political and social life. A few years later, after thorough preparation and preliminary studies, the People's Republic of China started to reform its economy and society, while "restructuring" began in the Soviet Union after 1985, under the initiative and leadership of Mikhail Gorbachev.

Whereas the Hungarian reform was introduced at a time of relative political consensus (despite the existence of varying expectations concerning the further development, expansion and significance of the reform), in China and the Soviet Union economic reform had to be preceded by the rearrangement of political power relations. In this respect, the position of the Chinese leadership was somewhat easier, because the extreme actions of the Cultural Revolution had destroyed the traditional power structure, thus providing immediate room for reform. Clearly, then, the momentum of socioeconomic reform is not influenced by its goals alone but by the internal political situations within which these goals can be realized.

Needless to say, the cyclical movement of socioeconomic processes must be taken into consideration in every case. On the one hand, resistance, attempts at slowdowns and passivity occur. On the other, tensions created by the reform itself arise—tensions which may ultimately change the power relations between the representatives of progressive and conservative views. In addition, many novel problems caused by the introduction of reform (e.g., shortages of goods, bureaucracy, slow reaction) do not appear in the traditional sphere, which gives many people the impression that these problems are alien to socialism and demonstrate the distortion of the political concept. Finally, it must also be taken into account that the world economic crisis has brought about an extremely difficult external economic environment, causing the results of the reforms to appear gradually. Nonetheless, it is also true that every government considers the reform of a well-functioning economy or society to be superfluous. It is a reasonable assumption then that, as a result of these meditations, we cannot come to any other conclusion than that which the conservatives have arrived at earlier in every society.

When one looks again at the Soviet and Chinese reforms (which have tremendous importance for world politics), one must recognize that economic growth, for which reform is a precondition, is the most decisive problem facing socialist society and the socialist system today. It represents the revision of earlier views on security questions, which increases the importance of the reform from the perspective of international politics, since the new approach offers new opportunities.

Finally, I should like to point out that the resolution and determination of the Soviet Union in the introduction of reform will, presumably, speed up the reform process in those countries in which the political leadership has insisted on maintaining the old system of control. The phasing out of the political leadership with a more conservative attitude toward reform may again result in the appearance of novel types of reforms, the characteristics of which shall be examined later.

Internal Reform and External Response

It is obvious from the findings and references above that two systems of requirements are mixed up in the reform movement of today: the replacement of the centralized-bureaucratic transformation model of the socialist economy and society by a modern model—one which promotes intensive development and leaves more room for principles of economic rationality—and adjustment to the new international economic and political conditions brought about by the epochal change in the world economy. (In the case of the Hungarian reform these two paths can still be clearly distinguished.) Of course, these two systems of requirements are not contradictory, since they mutually influence and complement each other; it is nevertheless necessary to point out that this situation decisively determines the quantity of changes and the speed of their adoption. Adjustment to the world economic situation has a great many elements of constraint. It is obvious, for instance, that the liquidation of weak enterprises must be sped up, that subsidies must be reduced and prices raised (since the load-bearing capacity of the budget is limited), that the regrouping of manpower must be accelerated, and that the reduction of the standard of living of some strata must be tolerated. Additionally, stimulation of productivity must be intensified; greater differentiation must be permitted in the distribution of incomes, allowing small enterprises and the private sector to enjoy an unjustifiably high income due to the put-off demand.

Some of the reformers harbored the fear that the problems and difficulties caused by the reform would be indistinguishable from those economic problems which arise independently, or which are the results of activities which impeded the introduction of the reform. Unfortunately this fear has proved accurate. It is unrealistic, however, to assume or to advocate that thorough socioeconomic reforms can be introduced in a vacuum.

There are a number of questions and problems which decisively influence the external economic policy of the socialist countries in the new economic way of thinking fostered by the reform. Reference has already been made to the fact that the external economic concept of the centralized bureaucratic transformation model turned inward in the 1950s and 1960s, due to the embargo, blockade and cold war affecting the socialist economies. One of the leading ideas of the model of the period of transformation was the avoidance of external economic dependence; that is, the socialist countries should not get into an economic situation (or only one of a limited extent and time) in which their internal concepts of economic development would depend on their cold war adversaries. The solution to avoiding dependence was industrialization, as a result of which key sectors of the economy would be able to produce for themselves all necessary goods. But this concept failed, as evidenced by the constant growth of imports needed by the different economies; the strength of a given economy is not provided by import restrictions but by the ability to export. Conversely, industrialization concentrating on the end products created a huge demand for intermediate components, such as sub-assemblies and semi-finished products, which in most cases had to be met by imports, since both the domestic supply industries and the division of labor within the CMEA were weak in this respect. Protectionist economic policy did not give high priority to building up adequate export capacities, and factories preferred to sell to markets with which they were familiar and which were less demanding. Consequently, the balance-of-payments problems of these economies are not the results of imports but of too few exports; i.e., only the development of adequate export capacities can provide a satisfactory solution, especially in the longer term. A new external economic concept is therefore needed, the development of which would have been inevitable even if no epochal change or crisis had occurred in the world economy.

The epochal change in the world economy makes it obvious, however, that the reforms must rely on intensive economic cooperation and adaptation to the world economy. This statement applies to both the individual economies and to the CMEA. Such steps have already been made in almost every European socialist country to varying degrees.

Recommendations

Of these changes—some of which occur within the framework of general economic reform, while others are being realized even within the countries adhering to a conservative economic policy—one should note the most important ones:

- The Hungarian economy explicitly accepts, and other economies accept in practice, the requirement of *export orientation*—i.e., the view that the growth of the economy depends primarily on the *ability to export*. This means in practice that in development policy, priority must be given to those economic activities which concentrate on convertible currency markets.

- In the course of the reform to be brought about in the CMEA, priority must be given to those activities and those forms of cooperation which are directed at third markets.

- The methods of industrial and commercial cooperation with the convertible currency countries must be updated, and in the course of their development the exporting of components, subassemblies, semi-finished products, and so forth must be accepted.

- The quality of our existing export products must be improved by the application of advanced technology in such a way that instead of a price competition they should be able to participate in a quality competition.

- Technology imports which further the development of domestic production, tertiary services, and the infrastructure must be increased.

- The foundation of joint ventures which provide technology and markets for the economy must be made possible.

- A flexible trade policy must be conducted which promotes the appearance of our products and services in new markets, adjusted to the requirements of the countries and markets in question.

In the above sketch of the reform policy and new external economic policy of the socialist countries, it has been assumed that these tendencies within the cyclical nature of social movements will continue, since rapid technical progress presumes flexible socioeconomic structures. In contrast, however, it is clear that the disruption of interconnected economic and production conditions, as well as factors of the world economic situation such as the extension of markets in circumstances of indebtedness, demand new harmonized steps of economic policy.

Furthermore, global factors that can only be resolved on a cooperative international scale must be taken into account, while the absence of cooperation and settlement endangers the survival of mankind and radically increases the possibility of potential conflicts. Finally, the system of interdependence that has evolved in the world economy, which determines the components of a worldwide business cycle, must be interpreted as a new phenomenon. Mankind is not yet able to handle these extremely important problems, since 180 national economies still exist and there is little hope that within the foreseeable future a supreme international power will be established which would be able to solve these problems institutionally.

The East-West Dimension

Unilateral readiness on the part of the European socialist countries for an opening towards the external economy is by itself insufficient. It is true that international conditions provide arguments for the renewal of East-West economic cooperation; yet two factors must be pointed out. The first is that economic cooperation is the result of common economic interests. If such common interests do not exist, cooperation cannot be brought about despite the best intentions. In this context, it must be stressed that in addition to existing common interests there are also potential shared interests, and their discovery and development are no less valuable than the technical innovation which furthers the solution of a given problem. Moreover, interests are never of a static nature, because existing contacts may represent the foundation and the driving force for new contacts.

The second important factor is that in the course of the historical and economic changes which have occurred since the 1930s, economic thinking has become strongly dependent on political security considerations, and economic considerations do not prevail on their own, even if such a possibility exists. Since subordination of the economy to politics has been a factor for more than two or three generations, today we consider economic points of departure which have become assimilated into the economy from politics over the past fifty to sixty years as both the natural and exclusive possibilities. From this we can only conclude that in thinking through the economic possibilities, the current and future political and security situation should be the starting point, not the prevailing postulates of the past fifty to sixty years. It is obvious that the reorganization of socialist society and the reform of the socialist economy will also involve changing some existing conditions. It is conceivable, for instance, that the quantity of foreign exchange available for purchases will diminish temporarily. (Consider China, which had substantial foreign exchange reserves before opening but has since accumulated debts in just a few years. This may represent temporary problems, but it would be ridiculous to claim that the depletion of the foreign exchange reserves sets back trade opportunities when a market made up of one billion persons and considerable capacities, which has been closed until now, is opening up.)

East-West trade expanded rapidly in the 1960s, but this occurred on an extensive basis, and the driving forces of that expansion have since become exhausted. *Stagnation* occurred in the 1980–1983 period, partly due to debt accumulation on the part of socialist nations; the driving forces of the intensive stage had not yet unfolded.

Therefore, in 1983–1984 East-West exchange represented only 2 percent of world trade. According to an international survey, the share of East-West trade in the foreign trade of the CMEA countries has also diminished since 1980, presumably due to the forced reduction of imports caused by tensions in the equilibrium. If we project today's trends into the future—

assuming an unchanged course of economic policy—we cannot count on a substantial improvement in the coming years: the expansion of the export capacities of the CMEA countries (which is linked to restructuring) is a relatively slow process, and consequently, in order to ensure the balance of payments, imports cannot be raised to the necessary level.

Prospects for European Cooperation

The loss of economic ground by the whole of Europe is a long-term, large-scale process. In the first quarter-century after World War II, the economic dynamism of Eastern and Western Europe far surpassed the world average, but in the 1980s it has lagged considerably behind Japan, Southeast Asia and the United States. This falling behind has acquired such clear and characteristic forms that we are justified in speaking of an end to the Atlantic period and the emergence of a Pacific period. Compared to the 1970s, Europe's share in world exports and imports has fallen considerably, as well as in world engineering exports and imports, U.S. and Japanese imports, and even OECD imports.

Before World War I, European scientific and technical achievements were the foundation of the industrial development of the United States and Japan. In Western Europe, the deterioration of competitiveness on the world market is also due to the proportionally lower number of researchers employed in high technologies, the lower amount of money spent on research, the slowdown of growth, considerable unemployment, and the slow expansion of the internal market. The lower level of dynamism of Western Europe is attributed by numerous researchers to the "fragmented" nature of Western Europe compared to the United States or Japan; one might add the splitting of Europe into two parts, which came about in the 1940s, as a similar factor. However, broad economic cooperation and an intensive division of labor may soon become possible between Western and Eastern Europe through approaches to—and agreements on—military questions.

The reduction of trade and economic relations between East and West in the 1980s is the result of numerous new factors; they cannot be enumerated here, since there has as yet been no precedent for such discussions of economic policy among the Common Market, The European Free Trade Association (EFTA) and the CMEA. However, we can assume that the fall is not due to prices and the forced restriction of imports alone but is also attributable to structural and contact problems. In the exports of the CMEA countries, the overall share of industrial products is low (58 percent); in the case of the Soviet Union, it is as low as 12 percent. Approximately 80 percent of the exports of the CMEA countries are raw-material-intensive products or those which are sensitive to cyclical fluctuations. For the Soviet Union and Romania, the high price of primary fuels undoubtedly hid the weaknesses of the export structure.

On the other hand, from the geographic-historic side, East-West contacts were developed with an emphasis on Europe. The West European share of East-West trade was 82 percent between 1982 and 1984, the share of the United States and Canada was 8 percent, and Japan's was also 8 percent. There is no doubt that opportunities for developing economic relations are limited in both the old structure and the present monetary situation. This is one reason why, especially within Europe, the potential offered by an intensive division of labor should be thoroughly considered.

In the course of (or parallel to) negotiations between the Common Market (EC) and the CMEA, there would be a great need to analyze this potential. Some fundamental possibilities of a political nature should be examined:

- Should it not be possible—assuming a military agreement—to restore the economic unity of Europe? In practical terms, what kind of cooperation in economic policy and external economic relations between the existing organizations is possible?

- How would an intensive intra-European division of labor influence the competitiveness of the participants on the world market?

- What would the market of one party offer to the other party in the conditions of today's merciless competition?

In addition, one should also consider that the politics of the three biggest powers of the contemporary world are becoming increasingly centered on economics. The Soviet Union and the People's Republic of China have declared this to be the case, and economic reform is decisive proof of this intention. To all those acquainted with the nature of the economic reforms and able to foresee the complications involved, it is clear that at least one or two decades are needed for the implementation of such reforms. The political equilibrium evolving during the course of the transformation assumes the maximum concentration of the attention and potency of the leadership. Additionally, a reform of this kind channels the power and energy of national public opinion towards the transformation. For the time being, the United States has not declared similar intentions, but it is clear to the expert economist that while the model of overarmament may have in the short run some advantages—it may equalize other unbalances and strengthen the domestic economic position of some individual states—in the longer run the increase of the budget deficit and the balance of payments deficit will lead to the accumulation of debts and the weakening of the standing of the key currency. The U.S. economic-political situation confirms this finding in every respect; the peculiar thing is only that the lag—especially compared to Japan—applies to high technology as well. U.S. Treasury economists and policy-makers recite wonderful odds about the immeasurable advantages of the market economy, yet a considerable number of large American companies are dependent on the government treasury and weakened by supplies costs—phenomena common in the monopolistic state sector. Consequently, I believe that the United States—irrespective of the results of the 1988 elections—must in the near future replace overarmament with an economy-oriented policy.

An economy-centered development model adopted by the world's three leading powers fundamentally changes the international economic climate, since the evolution of world economic cooperation has often been rendered difficult by the unfavorable climate caused by political animosities between them.

Conclusions

At the time of this historic change in the world economy and world politics, the relationship between the political, military and economic factors should be carefully considered, including their movements, interaction and opportunities for conversion. An agreement in the military domain is a precondition for political and economic cooperation in our dangerous contemporary world. Conversely, the relatively narrow role of the military factors in everyday life should be taken into consideration; these factors do not bring about a mutual interest and are not built organically into the system of actions of the national society and economy. It is obvious that international verification systems will be established, that the signatories to the various agreements will control each other, that it is necessary to consult in the case of the appearance of new technical trends, and so forth, but this is generally the problem of a narrow stratum of experts, who will have sufficient information and grasp of the different alternative solutions.

In contrast, economic interests are built deeply into the life and fate of any national society, since they represent workplaces, income, profit, growth and new markets for the enterprises and national economies participating in the cooperation. In addition, they also determine in the longer run the international positions of the different countries. Therefore, it is hardly possible that some agreement on the restriction of armaments can prevail on its own without common goals—or, particularly, without economic interests.

Many claim that the balance of power forms the main foundation of the maintenance of peace, if contrary or diverging power interests are assumed. But what do we mean by the balance of power in the last decades of this century? General balance, or a balance of all specifics of power? Worldwide balance, or a separate balance with respect to every region? Can the various specifics of power be converted; does the preponderance of a particular party in one region neutralize the preponderance of another party in another region? Can it be considered a balance if the situation is balanced militarily but one party has overwhelming economic power, or vice versa? These are, of course, questions which can only be answered by science, particularly a multidisciplinary science, over the long term.

Unfortunately, in the course of the necessary cooperation between science and the political-economic leadership, it is the researchers who adjust to the short-term reference of the political-economic problems, and not the political-economic leadership that learns to practice the more complex, long-term thinking that is crucial at this stage.

Of course, problems of this nature cannot be solved overnight, but at a time of such fundamental changes science must do everything in its power to approach the new problems with new and more perfect methods.

9

The Impact of Technology and Technological Change on Communications, Culture and Public Policy Issues

PEKKA TARJANNE AND MAURI K. ELOVAINIO

In our information society, the old saying "knowledge is power" (Francis Bacon) should be rephrased: "knowledge, informatics and the transfer of information (communications and media) create power." Correspondingly, the substance and the structure of the "Fourth Estate" have widened greatly. Compared to other contributory factors of power, the media are the most difficult to control in the world, "shrunk" by the unexpectedly rapid development of communications technology.

Introduction

In our modern society and especially in highly industrialized parts of the world, social changes affecting even the basic infrastructure have been profound, the primary cause being rapid scientific and technological development, particularly in the field of information technology. Consequently, the last decade has been both politically and societally very challenging for decision-makers dealing with the impact of ever-accelerating technological development.

It is obvious that the societal impact of technology—in particular, information technology—is twofold. On the one hand, technology provides the opportunity to accelerate technical development and structural change in welfare states of the

Western type; on the other hand, it also creates unplanned structural changes, raising several severe unemployment problems and leading to efforts to guarantee as high a rate of employment as possible. But one can argue that where the strategic advantages of information are disregarded, even more difficult problems are likely to arise concerning employment. Knowledge, in general, is a major component of production know-how. With the help of telecommunications, one can master this essential factor of production. Because of increased knowledge, it is legitimate to use the term "information society" for this phase of development in modern society. Knowledge, as well as other factors of production, must be used efficiently. It must be mastered; it should not be allowed to lag behind the times and become inaccurate, or likewise to disappear. One must pass knowledge on in order to work efficiently. Information technology makes information available much faster than other methods.

It is widely held that in general Western Europe contributes about 30 percent, the United States about 40 percent, Japan 15 percent, and the Soviet Union about 15 percent to the input of research and development toward new knowledge. This substantiates the general observation that the countries of "the Pacific Axis," i.e., the United States and Japan, contribute the largest share. High-technology product development and industrial processing in high technology of these countries suggests the same tendency. Correspondingly, when inputs are compared to the diffusion of the products in the world, one perceives that the input/output relation of research and development expenses is very slight on Europe's part. In reality, one must further note that Europe does not lag behind the United States and Japan as far as the level of knowledge or technology is concerned, but only as regards the industrial and commercial exploitation of this knowledge. However, Europe has during the past few years clearly recognized the reality concerning technological development and its application to commercial productivity.

This study concentrates on telecommunications in its wide spectrum, and mentions the problems related to mass communication only in passing. Nevertheless, it is self-evident that whenever the problems of communications as a whole are discussed, and particularly when emphasizing contemporary East-West relations, the ideological elements within transnational commercialized broadcasting are crucial to the national interests of the various countries belonging to the various politico-ideological blocs in existence. Referring to the motto of this paper, in the name of the protection of one's own national integrity and cultural image it must be acceptable to select and possibly to control the contextual influence of transnational mass communication.

In principle, the whole world, and Europe in particular, shares a considerable mutual understanding about cultural problems that may arise, for example, because of supranational commercial TV broadcasting or the use of the "electronic media" for purposes of propaganda. At the meeting of communications ministers of the Council of Europe in Vienna in December 1986 (Finland is not a member, but attended the meeting with full authority), it was clear from the speeches delivered by both Finland and all the other participating countries that there is a need to establish basic international rules to control program broadcasting that crosses national borders—in other words, to create some kind of minimum regulation code. In a sense this problem is analogous to the vast set of problems in the 1970s involving the control of transnational corporations. At that time, "rules of the game" were created within the framework of the United Nations.

The root of this problem is that no one denies the importance of control to protect and develop our national cultures, but at the same time international public opinion demands the opposite of protection: liberalization, deregulation and privatization in all the essential fields of communications. Thus the questions concerning the control of communications contradict in various ways the above-mentioned "neoprinciples." This has already been seen in the West European experience in applying these principles. Even though the fundamental problem is subtle and internationally valid, it does not provide any possibilities for quick and easy solutions.

West European Technological Development

The division of Europe into two different ideological, economic and security entities is today a reality, despite several current "detente-favoring" factors positively influencing the existing cleavages between these two blocs. The rapid technological development occurring in both Eastern and Western Europe and new advances in information technology and telecommunications have had a great impact on the social structure of both types of societies.

In the development of the modern technological society, trade, political issues and security problems are deeply intertwined. Clear lines between these different domains are gradually diminishing. In all West European countries, enlarged investments in technological development have essentially followed one and the same pattern, intensifying particular industrial efforts in high technology. The high level of competition with the United States and Japan is stimulating a technological renaissance in Europe today.

This technological renaissance is evident within the European Community (EC), with a large variety of special technological programs (e.g., Esprit, Race, Brite); and within the European Free Trade Association (EFTA) countries, which are tightening their mutual cooperation and at the same time intensifying their relations to EC development programs. In Western Europe the convergence of different technological sectors is gradually accelerating; within this process, the role of information technology, especially telecommunications (the convergence between these two sectors is growing and will restructure the markets), is central not only for communications but for overall industrial development. Within the European Community it has been estimated that before the end of this century the telecommunications industry will be the largest and most important sector of industrial production. In 1986 the telecommunications share of GDP of the EC countries was 2 percent, but this share is estimated to reach as high as 7 percent by the turn of the century.

But will this European renaissance be efficient enough? Or is the gap between Europe and the leading countries, i.e., the United States and Japan, still widening? The latest important invention, the so-called room-temperature superconductor, is predicted to cause real revolutions in many fields, ranging from loss-free power transmission to cheap and rapid levitating trains and from medical instrumentation to superconducting supercomputers. But the experience of the "superrace" of the past eighteen months seems to show that the winners in basic research are the Americans and the winners in economic applications are the Japanese. Europe, in order to keep from falling off the wagon altogether, needs much stronger joint efforts than we are used to today.

When one analyzes the integrative technological development and the integrating tendencies of telecommunications in many of the EC and EFTA countries, one has to mention some of the many programs and organizations promoting the common goal of technological integration. First of all, one must consider the joint Eureka cooperation, the launching of which has been very rapid and which will encourage numerous projects dealing with information technology and teletechnics. CERN, ESA, COST, Eurodata, among others, have an important role to play for common West European technological interests. In this connection one should not omit the bridge-building role of several organizations of different industries in both EC and EFTA countries (UNICE, ECTEL, Roundtable of European Industrialists, and so forth).

In the field of telecommunications, the role of the European Conference of Posts and Telecommunications Administrations (CEPT), composed of 26 Western European PTT administrations, is especially important to the creation and acceptance of new common standards for Western Europe. CEPT provides the EC and EFTA countries with a forum for discussing and drafting future outlines for telecommunications policy. Recently, industry representatives have also entered into the work of CEPT, and they now participate in the activities of the numerous working and subworking groups. In this context one should emphasize the role of satellite organizations such as EUTELSAT and the role of the worldwide INTELSAT (Socialist countries have Intersputnik) and INMARSAT in the fields of telecommunications, particularly in mass communication. The West European countries naturally also play a very important role in the International Telecommunications Union and in its several specific bodies.

The difficult yet crucial issue in East-West relations today is that our knowledge concerning the tele-technological development in the socialist countries, and especially in the Soviet Union, is scarce. However, as far as we know, important decisions were made by the Soviet leaders in early 1987 to greatly increase investment in telecommunications technology and telephone systems. The principle of openness and democratization as well as the striving for efficiency in Soviet production can be seen as indications of concrete changes in the technological infrastructure.

Looking at the future of Europe and keeping in mind the idea of genuine pan-Europeanism, as well as recalling the aims of the CSCE to promote security and cooperation in the whole of Europe, it is highly relevant to speculate whether actual development in technology and telecommunications will unite or separate the two Europes. In other words, will development result in "technological" isolation of the socialist countries in Europe, and in the deterioration of the present telecommunications situation? The best channels for human contacts, both in business and between ordinary people, are well-planned and accurate telecommunications networks. It is no doubt clear that the possible deepening of the economic and technological "gap" might indirectly have an effect on the development of Europe's security policies and, in any case, damage the cooperative spirit that now prevails in Europe.

Future Prospects for Global Information and Telecommunications Systems

The Importance of Fiber Optic Technology

Today's telecommunications technology is becoming more and more integrated. Thus the same technology—often, even the same equipment—will serve both targeted and mass communication in the future. Also, various types of networks will gradually be integrated into one single unit. The telecommunications networks have already been integrated to such an extent that sound, image, text and data are transmitted within the same network. This highly integrated telecommunications network will develop even further into the Integrated Services Digital Network (ISDN). On the path to the ISDN, optical cable plays an essential role.

Fiber optic technology is one answer to the increasing demand for transmission capability. The demand for fiber optics grew out of the need to send large quantities of information over immense distances at minimum cost. In the coming years, fiber optic systems are expected to become the most inexpensive cable system available. This is due to the predicted decrease in fiber optic cable prices and installation and maintenance costs.

At higher transmission capabilities, fiber optics are even more attractive. They offer both broader band width and long repeater spacings. Fiber optic cable technology is already used on a minor scale in the telecommunications networks, but it is important to note that optical technology also has begun to be used in transatlantic submarine cable connections.

The development of fiber optics will make it possible to integrate transmission in different telecommunications networks.

Fiber optic cables will permit expansion of the range of digitally based services and provide the infrastructure necessary to implement the ISDN.

Optical cable will be the infrastructure of the future global telecommunications network. Satellites, for instance, serve well for some purposes; however, despite the rapid technical development of communications satellites, there are certain specific problems connected with them; for example, the limited area dedicated to communications satellites in space is rapidly being filled. If one looks at the backbone of the telecommunications network, there is no doubt that the terrestrial optical cable network will be the most important factor in the development of telecommunications systems.

The Integrated Services Digital Network (ISDN)

The efficiency and economy of the applications of new technology are the most essential issues in all telecommunications organizations today. Teletechnology is developing at a constantly accelerating speed towards a continually higher grade of complexity. For the time being, one of the most discussed subjects in the international telecommunications services is the ISDN.

On the one hand, technology—that is, fiber optics—provides new possibilities. On the other hand, consumers develop new needs. The development of activities is based on consumer demand for communications. In the course of technological progress, more and more new services are made available.

The modern digital technology employed in the construction of the telephone network forms the foundation of the integrated services digital network. The availability of digital circuits is a major precursor to the introduction of the ISDN, which will benefit users by providing a single standard access to a wide range of telecommunications services.

The construction of the ISDN network is a major objective in almost all the developed countries, and many of them have already started trial operations. As early as 1984, the International Telecommunications Union's (ITU) International Telegraph and Telephone Consultative Committee (CCITT) had already defined international standards for the basic characteristics of ISDN.

Within Western Europe, the planning and development of ISDN comprise today's reality in telecommunications. In the

United Kingdom, ISDN trials have taken place since 1985. In France, the first ISDN trial network was begun in the fall of 1987, and in Finland in early 1988. In the Federal Republic of Germany, where the ISDN is seen as a stepping stone to the development of switched broad-band services, the first two ISDN exchanges are entering operation.

The development of networks is, in fact, ultimately leading to the broadband ISDN, on which one can base future services, many of which have yet to be invented. In the past ISDN was just a vision widely talked about by specialists; today it is beginning to become a reality.

The prevailing enthusiasm for ISDN on the part of telecommunications people must be restrained by the fact that before ISDN becomes viable, there are still many problems to be resolved; in particular, one can point to the existence of incompatible national prestandards, incompatible signaling systems, different introduction dates, and lack of agreement on numbering, tariffs and terminals. The price of terminals must go down, which in turn depends on suitable very low-speed integrated circuitry. There is a lack of equipment tailored to ISDN operation and no agreed upon CCITT standard to which that equipment might be made.

There is very seldom enthusiasm without some criticism. Many potential users feel some suspicion about ISDN—quite normal when new technology not yet in use is introduced to the public. The most serious criticism of ISDN plans have been that far from serving the interest of users, they are being used by PTTs to shore up their crumbling monopolies. In the private sector, it has been argued that PTTs should not use ISDN as an excuse to reduce the availability of private circuits. In the future world of ISDN, it will be important to keep open the option of user networks tailored to the individual requirements of companies.

A Pan-European Integrated Mobile Telephone Network—A Reality?

The creation of an integrated European mobile telephone system has long been discussed in high-level negotiations between European organizations—political groupings as well as those focused on telecommunications cooperation and administration. An historic common decision concerning an international mobile telephone system was reached in February 1987, during the joint meeting of European telecommunications administrations on Madeira.

As a result of the Madeira decision and the May 1987 Bonn agreement, during the next decade Western Europe will get an integrated mobile telephone network which will, as a new international digital network, be a parallel system to the national analog systems. For some West European countries this is the first and only real cellular mobile telephone network. There is a common agreement that the narrow-band system will be the basis of the West European international system. Thus, in the near future an integrated mobile telephone network will extend from the Nordic countries to the southern parts of Europe, on the condition that the agreement reached will be carried out in practice. By mutual decision of the West European countries, some kind of test system should be introduced by 1991.

This decision also puts the European mobile telephone producers in a better position in comparison with their Japanese and American counterparts. Nokia-Mobira of Finland was asked for suggestions when the basis of the system was being laid out. Mobira, like most of the other counterparts, favored the narrow-band network that was ultimately chosen. When the network is ready, it will be possible to use the new digital mobile telephone in all parts of Western Europe. This means in practice a "complete delocalization" of the telephone, because not only fixed car phones or portable telephones but also small hand-held portable telephones will be able to be used everywhere within the network.

The agreement will probably also lead to mass markets selling pan-European mobile telephones. Nokia-Mobira has already signed an agreement on cooperation with Matra of France. Other European companies interested in mobile telephones include Siemens, Bosch, Alcatel, Philips, Ericsson, GEC, Racal and Plessey.

This agreement, which has been approved by all the West European countries, is politically significant; this step forward in teleindustrial policy is a very important achievement, following in the pattern of the general development of West European technological cooperation discussed earlier.

The Role of the Nordic Countries

The Nordic countries have long enjoyed excellent cooperation in telecommunications. They have an agreement on the construction of a joint network, which can be considered unique; its fundamental idea is that a telecommunications network, with connections both within the Nordic countries and outside, will be designed, constructed and used jointly. In this way the Nordic countries can save, as a whole, some ten million ECUs a year.

As a result of the Nordic cooperation, both the Nordic Mobile Telephone (NMT) network and the Nordic Public Data Network (NPDN) were created. NPDN is the largest integrated data transmission network in the world.

Within the cellular NMT network there are two systems in operation: first, the NMT 450, which was introduced into service five years ago, and second, the NMT 900, introduced this year. The NMT system was developed and the network constructed jointly by the Nordic telecommunications administrations, whereas the trade of mobile telephones is open to free competition. In this way it is possible to guarantee consumers the highest possible service standard: fairly low prices, wide choice of equipment and rapid product development. The fact that the Nordic countries have carried out this project together right from the planning stage has ensured an economical use of the limited resources of radio frequencies. Moreover, this cooperation has made it possible to provide sparsely populated areas with reasonable coverage by the network.

The new NMT 900 network was developed to alleviate heavy traffic. At the moment it has over ten thousand subscribers. The NMT 900 was designed and constructed in such a way that in addition to fixed car phones and portable telephones, even handheld portable telephones can be used.

The innovation and construction of the two automatic mobile telephone networks has provided the Nordic telecommunications administrations not only with the leading position in the world in the mobile telephone field, but also with an expertise and knowledge for which there is worldwide demand.

Telecommunications as a Consumer Service in Europe

Within Europe, one could say that the whole society is somehow connected to telecommunications systems. The present telecommunications technology affects all of us, and in the future the impact of teletechnology will certainly become even more pronounced in human contacts in different circumstances: in ordinary human life, industry and business.

Industry, technology and trade imply internationalism, and this term in turn always includes some kind of transnational action. Telecommunications make fast contacts and communications possible for European business life, which plays an essential role in turning the wheels of industry, thus providing the entire foundation for the development of new technology.

A certain paradox lies in the fact that we are still living in the political framework of individual national states, especially in terms of protection of national cultural identity. At the same time, business life has become thoroughly internationalized and telecommunications have, consequently, achieved their own international identity. Individual states, when they perceive their national interests to be threatened, feel that they are only defending their legitimate constitutional rights when they put obstacles in the path of the international development of telecommunications systems, and this may jeopardize the telecommunicational interests of the business community.

Liberalization, privatization and deregulation are, at least to some extent, favored by West European telecommunications administrations. The starting point for their services is consumer needs. The different, and to some extent tailor-made, telecommunications applications are very important to improving pan-European business communications services. Facing tough international competition, the European business community must have effective trans-European telecommunications to be able to maintain rapid connections with business partners. Modern telecommunications must provide the necessary services for the benefit of enterprises. Nothing can replace human contacts, but telecommunication services can make business contacts easier, faster and less expensive with all the new kinds of services, such as video conferences and other applied modern teletechnology.

Conclusion

In the future, perhaps the majority of telecommunications installations and various terminal equipment will function as mobile units. One can presume that mobility will be more the rule than the exception. The global telecommunications network is already the most intelligent apparatus in the world. One could compare the telecommunications network to a spider's web, which is at the same time becoming continually tighter and intertwined.

When looking at Europe as a whole, one might well ask whether telecommunications are really developing like that in practice. Actual developments in East-West relations indicate that some blank spots or borderlines, where communications do not progress, may be forming.

From a global point of view, the question of East-West relations is not broad enough; there must also be concern about the future of humanity. The North-South dimension must also be kept in mind. The "Missing Link" report published a few years ago by the Maitland Commission painted a gloomy picture about the standard of telecommunications in developing countries: half of the world's population shares only ten million telephones, or, in other words, there are more telephones in Tokyo than in all of Africa with a population of 500 million. Moreover, the distribution of telephones within the developing countries is extremely uneven. In the majority of these countries, 70–80 percent of the population lives in the countryside, and agricultural products are clearly the most important source of foreign exchange. Nevertheless, usually less than 10 percent of the total number of the country's telephones are in the countryside, and the quality of telephone service is very poor. Of course, many more dramatic examples could easily be given. But this is not our subject today.

Against this worldwide background, the standard of services for those living in the European telecommunications environment is on a very high level. At present, the core of the European problem is not a lack of facilities for perfect telecommunications services; rather, it is the question of creating a political climate in which it is possible to establish a future pan-European telecommunications network which can later be enlarged into a global, transnational network.

IV
Technology and Policy

10

The Impact of Technology on the Future of European Security and Cooperation

KONRAD SEITZ

Two key new technologies, microelectronics and genetic engineering, are setting in motion not only a technological revolution but a social transformation. They are leading us out of the industrial society into a new society: the information society. Information—more precisely, the ability to use machines to process information—is becoming the most important factor of production.

This ongoing social transformation is most clearly evident in the United States, where the information society is furthest advanced. During the past fifteen years, the number of blue-collar workers in manufacturing has decreased from more than a third to less than a fifth of the American labor force. For the democracies of the industrial triangle (the United States, Western Europe and Japan) as a whole, economists forecast that by the year 2010 blue-collar workers will constitute no larger a proportion of the labor force than farmers do today—that is, one-twentieth of the total.

The age of the blue-collar worker is waning, that of the knowledge worker is dawning. This also means that in the Western democracies the age of the masses, the social democratic age, as Professor Dahlendorf has called it, is coming to an end, and a new liberal epoch is rising. Initiative, creativity, entrepreneurship at many levels, responsibility for one's work, decentralization, access to information both within a society and across national borders, reducing the steps on the ladder of hierarchy—these are the new keys to work and life in the information society.

In the East, Soviet General Secretary Gorbachev has drawn similar conclusions from the trends inherent in the new technologies. The key words of his new policy are *glasnost'* and democratization, and indeed these demands will have to be fulfilled and made reality to prevent the East from remaining stuck in the industrial age.

Thus if we assume that both West and East successfully master the transition into the information age, then two new, transformed societies will evolve. It is difficult to imagine that the political relations between these transformed societies would not be changed as well.

The new technologies are global—by nature and in several specific respects. First, as far as communications is concerned, the world will continue its trend toward becoming a "global village." Communication satellites, direct broadcast satellites and television satellites with several language channels will further "shrink" the world. A British royal wedding or an Olympic soccer game is already a global event. The list of such events will grow. We may see the emergence of international politicians pleading their causes directly to an expanded populace, encompassing many countries. It will be harder than ever for governments to separate their audiences, telling one thing to the domestic public and another to foreigners. To control the flow of information, to jam foreign radio stations, will become ever more difficult and ultimately ineffective. In any case, blocking information from the outside would be incompatible with building a competitive modern society.

Second, the new technologies will lead to a globalization of markets and production.

Within the Western industrial triangle of the United States, Western Europe and Japan, national economies are becoming increasingly integrated. One product is often composed of components from many countries. Whereas in the thirty years prior to 1980 international trade played the leading role in economic integration, the primary role has now been taken over by international investment. Cross-investment and joint ventures are multiplying within the industrial triangle. So far, joint ventures are mainly bilateral: European-American, American-Japanese, European-Japanese. But Kenichi Ohmae, a Japanese management consultant, is already advocating the idea of a high-technology triad. Cooperation of the future will come in form of the triad enterprise and, in particular, the triad consortium that combines firms from the three regions.

The main driving force behind this sort of integration is the high cost of research and development of the new technologies. These costs can only be recouped if firms, rather than selling the products only in their domestic markets, immediately enter the world market.

Eastern Europe is still a closed trading bloc. Its share of world trade is only 2 percent. But it is clear that its members must also integrate into the world economy if they want to participate in the evolution towards the information society. A single region can neither develop all the relevant new technologies nor offer a sufficiently large market for those high-cost technologies. General Secretary Gorbachev's new policy is going in the direction required by the new underlying global trends. Successful integration of these countries into world trade and world investment will contribute to the economic progress of the East in a major way; at the same time, it would be a boost to world trade as a whole. Thus, successful integration of Eastern Europe into the world economy could become a major engine of growth for the world as a whole.

Furthermore, there are tasks in science and technology of such a scale as to require the combined scientific and financial resources of both the West and the East: high energy physics, controlled nuclear fusion and, above all, exploration and utilization of outer space. East-West cooperation on such large-scale projects cannot but have a far-reaching effect on the overall political relationship between West and East.

The new technologies also create a need for a new scale of international cooperation in another sense: to protect the environment and to control and avert the dangers arising from some of these new technologies. The Chernobyl nuclear accident has dramatically shown us that radioactive pollution knows no boundaries and that nations have a vital interest not only in the safety of their own nuclear reactors but also that of their neighbors' reactors.

A requirement for common action and control will also arise out of the immense and new possibilities which genetic engineering opens up to us. We need common safety standards to prevent, for example, the release of new microorganisms into the environment without adequate prior examination and adequate controls. We also need common ethical and legal standards to prevent the misuse of genetic engineering in its application to human beings. In the distant future it may become possible to clone human beings; it is clear that we must take care in time to absolutely prevent any use of genetic engineering which would destroy human dignity.

The new technologies, in sum, create new and powerful motives and a new necessity for East-West cooperation. Peaceful coexistence of West and East is no longer enough; we need cooperative interdependence.

Large-scale economic and technological cooperation presupposes mutual trust. Mutual trust, in turn, can exist only if there is security and respect for human rights in the whole of Europe. In present-day Europe a sense of security is absent; instead there is insecurity and mistrust.

Why is this so? NATO and the Warsaw Pact both proclaim their defensive character. But only NATO translates that claim into its actual force structure. Its strategy is defense at the frontier, and NATO forces could not undertake an invasion even if they wanted to do so.

The Warsaw Pact—and this is the decisive difference—has a strategy of forward defense: in case of attack their forces will immediately carry the war into enemy territory by a large-scale invasion. To carry out this strategy, the Warsaw Pact has built up massive superiority in those weapons systems crucial for an invasion capacity: battle tanks, armed personnel carriers, and so forth. Unfortunately, such a force structure is not only very good for defense, but, as perceived in the West, it also provides a capacity for a blitzkrieg-type offensive. This is the problem.

The June 1987 East Berlin declaration of the Warsaw Pact on military doctrine proclaims the following goal:

> To reduce the forces and conventional armaments in Europe to a level at which each side can guarantee its own defense, but does not have the means for waging a surprise attack.

If we can reach this goal, then the road to large-scale cooperation in Europe indeed will be open.

11

Technology and Public Policy in East-West Relations: The Return of Politics

THOMAS W. SIMONS, JR.*

In this essay, I address the impact of technological change on public policy issues. I will focus on two effects of technological change which are already clear in the socioeconomic arena. The first is the shift in work-force distribution in all our developed countries, not only out of agriculture but also out of manufacturing, into services and especially information services. In another essay in this volume are figures showing that services in general now employ over half the U.S. work force and are likely to employ three-quarters within a decade. Within those figures, however, information services alone now employ half the U.S. work force and should employ two-thirds by the late 1990s. The cause is technological: manufacturing in the information field simply requires less material input than other types of manufacturing. But the effect is profound in every area.

The second effect has to do with the particular consequences of the first for transnational economic exchanges. Trade in goods is growing, but is much less important than it used to be. It is much exceeded in value by financial transfers, which increasingly drive world economic developments. Economic growth is increasingly divorced from the manufacture of things, and low labor costs are less and less significant as a fac-

---

*The views expressed here are personal and do not reflect the views of the Department of State.

tor of comparative advantage in economic exchanges. In another essay in this volume are figures showing that for pharmaceutical drugs and telecommunications equipment knowledge accounts for 50 to 70 percent of the cost, and labor for only 12 to 15 percent. It will not be long before labor costs are reduced to 15 to 20 percent of the cost of manufacturing in most developed countries.

The most profound consequences of these two effects will probably come in North-South relations. The outlook is for long-lasting marginalization of most of the Third and Fourth Worlds when it comes to purely economic factors. There are probably only 15 to 20 nations in the developing world which have the capacity to compete in such an environment, and that mainly by subcontracting for the developed countries. Dependence on the raw materials developing countries have traditionally supplied will be vastly reduced--probably only oil and water will remain as truly strategic commodities and this in turn will reduce strategic competition in the developing world, except to the extent the governments of the developed countries and their publics *care* about the Third and Fourth World tension and crisis situations that are likely to become endemic. As a result of technology, our publics are going to have better access to information, and they may well care—terrorism and the current crisis in the Persian Gulf are examples of such issues that concern the developed nations. Moreover, the escalatory potential of any use of weapons in a nuclear age means that these crises will continue to be dangerous. But developed countries will have less and less *economic* incentive to care, and the extent to which they do will depend more and more on political factors.

Still, this is a volume on East-West rather than North-South issues, and I would like to focus on the effect of the two changes I have identified on politics in our developed countries.

At the outset, I will express my personal opinion that the major opportunities and dangers which lie before us as a result of these changes do not have much to do with the military security equation. Of course they *could*; I have just pointed to the possibility that our public opinions and governments can *believe* their interests are vitally engaged in Third World crisis situations even without serious *economic* justification, and the same is true *a fortiori* for crises directly involving us. Since our countries are packed with weapons, the danger is obviously real, and important. But my own sense is that the information revolution reduces rather than increases the dangers of East-West military confrontation.

This is not so much because our publics are better educated and better informed. They could still be the electrically lighted barbarians Karl Krause talked about in Vienna in 1909. It is more because state control is weakening and scientific and technological complexity is increasing to the point where the lead time between research and development of new military systems and their deployment is extending rather than shrinking. The Soviet Union still has an advantage here, and that is the real reason for U.S. efforts to tighten up on the transfer of militarily significant new technology. But the lead time is extending everywhere, and the military establishments on all sides are going to be less and less ready to fight. They will always need more time to prepare.

This factor does make low-tech wars in the Third World more likely, and that is of course dangerous because of their escalatory potential. But the elites in all our countries are hyper-aware of that danger, and our publics are becoming increasingly aware of it, and—as I argued above—there are fewer economic reasons to care, to see the conflicts which could cause such wars in East-West terms. So on balance I believe the danger of military confrontation on the East-West axis is less.

The Political Basis for Policy

It therefore seems to me that the main East-West effects of the two changes I am focusing on have to do with the political basis for public policy.

All our countries have spent most of the postwar period provoking and managing the rural exodus, the pumping of our agricultural labor forces into new urban and industrial environments. This has been a historic task. In the first half of our century it was badly managed, to speak chastely, and this "bad management" produced not just massive politicization but political radicalism, both on the left and on the right, which contributed mightily to two world wars.

In the postwar period this transformation has been better handled, at least in terms of international stability. Though treated differently in different countries, everywhere it has been handled by strategies of depoliticization, by the economization of politics. In socialist Europe, politics were simply abolished. Political issues were reduced by force to economic issues. Where politics remained political, they were defined as almost purely international. There was strict control of the cities and of the industry into which millions of peasants were flooding. And there was a vast buildup of the state apparatus, fulfilling the dreams of 19th-century elites in these countries, who were after all engaged in the 19th-century task of building national power, by building iron, steel and later petrochemical industries. But politics as a clash of competing interests and values was abolished.

An economization of politics also took place in Western Europe. European unification using the vehicle of the Common Market did have important political aims. It was justified in terms of ending Franco-German rivalry, and giving West Germany a home. These were authentic aims. But in our terms the main task was to manage the destruction of the traditional peasantries, and the transition to an urban manufacturing civilization, in such a way and at a pace that would avoid political radicalization. And that was accomplished. The Common Agricultural Policy, which we Americans dislike so much, was a critical tool. In this sense Jean Monnet may have been the father of modern Western Europe, but Sicco Mansholt was the mother of modern democratic Western Europe.

In the United States the same transition was managed—if the word can even be used—in a different way. The American party system at the outset probably fit typologically somewhere between the party systems of southern and Eastern Europe. But the state was much weaker than anywhere in Europe, especially in economic management, and it was the market that mainly managed the transition. The result was the rapid emptying of the American countryside and the near-destruction of American cities, of what there was of an American urban civilization. In both Eastern and Western Europe, state control and the pace of change kept the social problem down on the farm; we transferred it to the cities. This situation is now stabilized. The countryside is almost empty of surplus labor, and the cities are being rebuilt. In the United States, too, the transition was effected without essential political radicalization. One casualty has been the American political party system, but that is mainly a problem for the next phase.

This phase of depoliticization also brought forth its own ideologies. The process was a bumpy one, and those of us who lived through it will smile somewhat grimly, since the first steps, in the late 1940s and early 1950s, were highly political. In the United States, McCarthyism was not a trivial phenomenon, nor was it merely international politics *in petto*. But as the phase progressed, in the late 1950s and 1960s, everyone adopted one or another version of economism. In the United States the ideology was "the end of ideology" itself. In Western Europe there were various forms of Butskellism. In Eastern Europe there was "goulash Communism," versions of "social contracts" between rulers and ruled. Most policy debates, everywhere, had to do with the economy. In retrospect, the exceptions—Suez, Vietnam, the East European crises—prove the rule.

This phase, I would argue, is now almost over. The working classes, and the more or less stable political situations built around the aspirations of new manufacturing working classes, are disappearing. They are in fact disappearing—as a result of technological change—on the very morrow of their appearance.

Continued economic growth now more and more depends on competitiveness using new technologies, and everywhere there is a divorce, more and more final, between manufacturing production and manufacturing employment. Production in manufacturing is growing in absolute terms and is constant in terms of national income; manufacturing employment is falling both absolutely and relatively.

All our countries have had educational revolutions opening up large opportunities for upward social mobility, but these have all been based on expectations of stability, of increased income and of increased dignity, if not for the individual then at least for families, generationally.

Now we all have a situation where overall growth is slowing, and there is increased stratification into a service work force and a high-technology management elite, private, state or some combination. And both the elite and the work force are less stable, less secure and better informed than in the previous phase.

In that situation political leadership depends more than ever on the capacity for achieving and maintaining public support from publics which are both less homogeneous and better informed. Vertical channels of communication, between rulers and ruled, are better than ever. But the basis for political consensus—the solidarity of interests that was perceived throughout much of the postwar period, and that was ultimately forged by the functional linkage between manufacturing production and manufacturing employment—is disappearing.

What new solidarity of interests will take its place? How will it be formed, and sustained?

Reasons for Cooperation

Just as technology has helped create the questions, it helps create the conditions for answering them.

In particular, it creates conditions for answering the question of why East and West should care about each other at all as the danger of direct military confrontation recedes, as the cheap-labor argument for economic cooperation evaporates, and as the East-West technological gap fails to narrow even if it does not widen.

The immediate prospect for all our countries is repoliticization, a "de-economization" of politics, a trend away from the economism which has marked the whole postwar era in every developed country.

Economistic politics are no longer plausible, because structurally our economies can no longer deliver the *stable* growth and sure prospect of *family* social advancement which economism promised.

Politics in every country will increasingly focus on the tension between economic efficiency, on the one hand, and manufacturing employment, on the other. And this is an issue involving values, freedom and equality, equity and justice. Competition over values will be made explicit, once again, in politics. For the first time since the 1930s, or, if one prefers, since the late 1940s and early 1950s, politics will be "about" values, or, in other words, about politics.

Technology itself pushes in the same direction, in two direct ways.

First, technological development is destroying the power of traditional political organizations, along with their mass bases. As Harlan Cleveland has noted, the state is diffusing at the top; it is leaking sideways, as other institutions take over many of its traditional functions; and it is opening up below, as a result of the educational revolution and new access by previously excluded groups taking advantage of new technologies. Political parties are also disintegrating. This process is furthest advanced in the United States, but it is also taking place in

Western Europe and in the East, where the demands of political control and economic efficiency in new technological conditions are in sharpest competition. What is left is a diffuse competition for diffuse support, and this can only be conducted as a competition about values. Images, not issues, will be the name of the game. To me at least, this is as clear in embryo under Mr. Gorbachev as it is in the evolution of the Labour Party in Britain, or the SPD in the Federal Republic of Germany, or as it has been under President Reagan.

Second, the obverse—the other side of the same coin—is that modern telecommunications are particularly apt vehicles for this kind of politics, this kind of debate. And if it is possible to try to keep central control of information, the cost is high not just in economic efficiency but also in political efficiency. The effort can delay the return of politics to politics, but not prevent it in the end.

But if a return to a politics of values is certain, its course and outcome are not. If politics is to be "repoliticized" after a long period of economistic drought, if it becomes once again a competition over values addressed to a diffuse and unstable population, we must expect to see and participate in a powerful surge of *traditional* values, especially where they have been forcibly suppressed.

The history of this century makes us all afraid of such values. After all, they captured mass support with terrible consequences at a critical stage of the *last* major socioeconomic transition in Europe, from agriculture to manufacturing, and the most modern technologies of the time were put in their service in the process. It is of course a cause for concern that the values we debate must include those which have been most suppressed or marginalized in all our countries throughout the postwar period: nationalism, nativism, traditional religious values—perhaps in new form, as the rise of neo-Protestantism not only in the United States but in countries like the Soviet Union and Romania suggests—ethnic and racial pride, and even chauvinism.

But I see no necessary reason why history should repeat itself. One of the facts we are dealing with is that the transition from manufacturing to information economies is different. We are not in the 1930s, or even the 1940s or 1950s, anymore. As Ivan Berend, President of the Hungarian Academy of Sciences, has noted in Helsinki, we should not fight the last war, either economically or politically.

We will all have to incorporate traditional values more in our shifting political consensuses. There will be odd alliances of conservatives and liberals and leftists, and "neo-" versions of each, within all our countries, and across the East-West divide. The specific consensus will vary by country. But it is by no means given that it will be retrograde or reactionary anywhere. Democratic values are very competitive in today's world. Moreover, they are often part and parcel of the repressed "tradition" that will gain new vigor as the postwar structures and surfaces bend and break. Since I am not a central European, I am quite confident that they can prevail in any serious competition over the long term. The new consensus in each country can—and, I believe, will—be democratic and, to use traditional "socialist" terminology, peace-loving.

And, best of all, the repoliticization of politics in our countries can create a new solidarity of interests between East and West that does not depend on the common danger of nuclear war—in a situation where that danger is receding—or on considerations of economic efficiency—in a situation where raw materials or cheap labor (physical *or* mental) provide dwindling incentives for East-West economic cooperation.

This is a solidarity of interests which goes beyond the political fallout of Chernobyl or cross-border pollution. Such issues are important, particularly because they provide an opportunity to develop models for institutional East-West management of common problems. Such problems themselves are too narrow to constitute a convincing political base for mutual East-West interest and interaction, to replace the danger of war or the appeal of comparative economic advantage.

The solidarity of interests I see emerging is rooted in fundamental problems facing all our countries, the consequences of the divorce between production and employment resulting from technological change. Debate over values, political competition, is an inescapable challenge for each of us, and to all of us together. It is worthy of our most serious and sober best efforts. It will, I am convinced, not just keep us together, but bring us closer together.

About the Authors

Ms. Mary Albon is Program Officer at the Institute for East-West Security Studies. She graduated from Harvard College in 1985 with a B.A. in History and earned a Masters of International Affairs in 1987 at Columbia University's School of International and Public Affairs and the W. Averell Harriman Institute for Advanced Study of the Soviet Union. Prior to joining the IEWSS staff, Ms. Albon was an intern at IEWSS and at the International Institute for Strategic Studies in London.

Dr. Jozsef Bognar is Director of the Institute for World Economics in Budapest and is Professor of Trade Economics at Karl Marx University of Economic Science. He has been a member of the Hungarian Academy of Sciences since 1965. He has served as Minister of Information, Mayor of Budapest, Minister of Internal Trade, President of the Institute for Cultural Relations, Chairman of the Scientific Council for World Economy, and Vice President of the Hungarian Economics Society. He is Editor-in-Chief of *Studies on Developing Countries, Acta Oeconomica*, and *Trends in World Economy*. His publications include *Planned Economy in Hungary: Achievements and Problems; Demand and Demand Analysis in Socialism;* and *Economic Policy and Planning in Developing Countries*. He was awarded the Hungarian State Prize in 1970.

Mr. Mauri K. Elovainio is Secretary General of the General Directorate of Posts and Telecommunications in Finland.

Dr. Andrzej Karkoszka is a Senior Research Fellow at the Polish Institute for International Affairs. He is currently on leave from the Polish Institute as a 1988–1989 Krupp Foundation Senior Associate at the Institute for East-West Security Studies. A consultant to the Polish Central Committee and Foreign Ministry, he has participated regularly in Pugwash Conferences on chemical weapons and European security, and has been a delegate to the Conference on Disarmament in Geneva (1984–86), to the preliminary consultations for the mutual force reduction negotiations in Vienna (1973) and to the UN General Assembly (1971). From 1977 to 1980 he was a Research Fellow at the Stockholm International Peace Research Institute, and during 1986–1987 a Resident Fellow at the Institute for East-West Security Studies.

Dr. Michael Kaser is Professional Fellow of St. Antony's College and Director of the Institute of Russian, Soviet and East European Studies, University of Oxford. He is Chairman of the Foreign and Commonwealth Office Wilton Park Academic Council, and is General Editor of the International Economic Association. His publications include *Economic History of Eastern Europe 1919–75; Soviet Policy for the 1980's; Health Care in the Soviet Union and Eastern Europe;* and *The New Economic Systems of Eastern Europe.*

Dr. Joachim Krause is a Research Associate at the Foundation for Science and Policy in Ebenhausen, the Federal Republic of Germany. During 1986–1987 he was a Resident Fellow at the Institute for East-West Security Studies in New York, where he wrote *Prospects for Conventional Arms Control in Europe* (IEWSS Occasional Paper # 8). In 1987–1988 he served on the West German delegation to the chemical disarmament talks in Geneva. He has written extensively on arms sales to the Third World as well as confidence-building measures and other issues related to conventional arms control.

Dr. F. Stephen Larrabee has been Vice President and Director of Studies at the Institute since 1983. He previously served as Director of the Soviet and East European Research Program of the Johns Hopkins School of Advanced International Studies and as Visiting Professor in the Department of Government at Cornell University. From 1977 to 1978 he was a Research Fellow at the Program for Science and International Affairs at Harvard University. Dr. Larrabee has published widely on U.S.-Soviet relations and East-West relations, including most recently "Gorbachev and the Soviet Military" (*Foreign Affairs,* Summer 1988) and "Eastern Europe: A Generational Change" (*Foreign Policy,* Spring 1988). He is the editor of the Institute's volume *The Two German States and European Security* (London and New York: Macmillan/St. Martin's Press, 1989).

Dr. Joseph Nye is Director of the Center for Science and International Affairs and Professor of Government at Harvard University. He has served as Deputy Under-Secretary of State for Security Assistance, Science and Technology. His most recent publications are *Nuclear Ethics; Hawks, Doves and Owls: An*

Agenda for Avoiding Nuclear War; and "How Does Arms Control Affect Risks of Nuclear War?" Dr. Nye is a member of the Academic Advisory Committee of the Institute for East-West Security Studies, and was co-chairman of the Institute's bipartisan Task Force on Soviet New Thinking in 1987.

Dr. Konrad Seitz is West German ambassador to India. Previously he served as the Director of Policy Planning in the Ministry of Foreign Affairs in Bonn. His main areas of interest include technology, East-West relations, and Third World issues. He has published on European-American relations, security and disarmament policy, high technology, SDI and North-South questions.

Thomas Simons is Deputy Assistant Secretary for European and Canadian Affairs at the Department of State in Washington. He has served as Deputy-Chief of the U.S. Mission in Bucharest, as Political Counsellor in London, and as Director of Soviet Affairs in the State Department. He has published widely on central and East European history and culture.

Dr. Eugene Skolnikoff is the Director of the Center for International Studies and Professor of Political Science at the Massachusetts Institute of Technology. He has served on the White House staff in the Science Adviser's Office in the Eisenhower and Kennedy administrations, and was a Senior Consultant to President Carter's Science Adviser. His research and teaching has focused on science and public policy, especially the interaction of science and technology and international affairs, covering a wide range of industrial, military, space, economic and futures issues. He is the author of *Visions of Apocalypse; End of Rebirth?; Science, Technology and American Foreign Policy;* and *International Imperatives of Technology,* plus numerous articles and book chapters.

Dr. Pekka Tarjanne is Director General of Posts and Telecommunications of Finland. Dr. Tarjanne has held positions of Professor of Theoretical Physics at the Universities of Oulu and Helsinki, and has taught at Berkeley, Wisconsin, Princeton and Cornell universities. He has published numerous articles on theoretical nuclear physics and elementary particle physics. Dr. Tarjanne has served as a member of the Finnish Parliament, President of the Finnish Liberal Party, Minister of Transport and Communications and a member of the Finnish

Foreign Affairs Committee. He is Chairman of Nordtel (Nordic Telecommunications Administration) and of the Supervisory Board of Televa (Communications), Vice-Chairman of the Supervisory Board of the Finnish Mining and Metal Industries, and a member of the Supervisory Board of Finnair (the national airline) and Postipankki (banking).

Professor Dr. Marek Thee is Senior Research Fellow at the International Peace Research Institute in Oslo and Editor of the *Bulletin of Peace Proposals*. He has written extensively on military technology, arms control and the impact of the arms race. His most recent publications include *Military Technology, Military Strategy and the Arms Race* (London: Croom Helm, and New York: St. Martin's Press, 1986).